Biotechnology and Materials Science

Biotechnology and Materials Science

Chemistry for the Future

Mary L. Good, Editor

Jacqueline K. Barton, Associate Editor

Rudy Baum, Assistant Editor

Ivars Peterson, Assistant Editor

Nancy Henderson, Assistant Editor

American Chemical Society
Washington, DC 1988

Library of Congress Cataloging-in-Publication Data

Biotechnology and materials science: chemistry for the future

Mary L. Good, editor, Jacqueline K. Barton, associate editor...[et al.].

p.   cm.

Includes index.

ISBN 0-8412-1472-7
ISBN 0-8412-1473-5 (pbk.)
1. Biotechnology. 2. Materials.

I. Good, Mary L., 1931-   . II. Barton, Jacqueline K.

TP248.2.B5512   1988
660'.6—dc19   88-14544
                            CIP

PRINTED IN THE UNITED STATES OF AMERICA

Second printing 1990

Contents

MARY L. GOOD is the Immediate Past President of the American Chemical Society and President of Engineered Materials Research, Allied-Signal Corporation. She completed a Bachelor of Science Degree in chemistry at the University of Central Arkansas and Masters and Ph.D. degrees in chemistry at the University of Arkansas.

Dr. Good served the Louisiana University system with distinction for many years. She then entered an industrial career, assuming her present position in 1985.

Dr. Good has been president of the Inorganic Division of the International Union of Pure and Applied Chemistry and a member of the boards of the Industrial Research Institute and the National Institute for Petroleum & Energy Research. She was a member of the ACS Board of Trustees from 1972 to 1980 and its chairman in 1978 and 1980.

Among Dr. Good's numerous honors are the Agnes Fay Morgan Research Award, the ACS Garvan Medal, and the American Institute of Chemists' Gold Medal. In 1982, she was named Industrial Scientist of the Year by *Industrial Research & Development* magazine. She is a recipient of several honorary degrees, including a doctorate from Duke University.

Her service to ACS, other professional organizations, and federal government units has been long and extensive. In the science policy areas, she recently was appointed by President Reagan to a second six-year term on the National Science Board. She has served on advisory panels for a host of organizations including the National Science Foundation, the National Institutes of Health, the National Bureau of Standards, the Office of Air Force Research, and Brookhaven and Oak Ridge national laboratories. She also was a member of a National Academy of Sciences panel on the impact of national security control on technology transfer.

Contributors

Philip H. Abelson
American Association for the
   Advancement of Science
Washington, DC 20005

Jacqueline K. Barton
Columbia University
Department of Chemistry
New York, New York 10027

Rudy Baum
Chemical & Engineering News
261 Capricorn Avenue
Oakland, CA 94611

Peter B. Dervan
Department of Chemistry
California Institute of Technology
Pasadena, California 91125

James Economy
IBM Corporation, K42-282
San Jose, California 95114

William A. Goddard, III
California Institute of Technology
A. A. Noyes Laboratory of Chemical
   Physics
Pasadena, California 91125

Mary L. Good
Allied-Signal Corporation
Des Plaines, IL 60017

Nancy Henderson
Science News Service
Washington, DC 20036

Emil Thomas Kaiser
Laboratory of Bioorganic Chemistry
Rockefeller University
New York, New York 10021-6399

Stephen J. Lippard
Department of Chemistry
Massachusetts Institute of Technology
Cambridge, Massachusetts 02139

Ivars Peterson
Science News Service
Washington, DC 20036

Gregory A. Petsko
Department of Chemistry
Massachusetts Institute of Technology
Cambridge, Massachusetts 02139

Howard E. Simmons, Jr.
E. I. du Pont de Nemours and
   Company
Wilmington, Delaware 19898

William Slichter
55 Van Doren Avenue
Chatham, NJ 07928

George M. Whitesides
Department of Chemistry
Harvard University
Cambridge, Massachusetts 02138

Mark S. Wrighton
Department of Chemistry, 6-335
Massachusetts Institute of Technology
Cambridge, Massachusetts 02139

Preface

*T*his book was inspired by a select conference on Advances in Biotechnology and Materials Science held by the American Chemical Society in early June 1987. The timing of this public forum was crucial because American science, the nation's competitiveness, and ultimately our standard of living may come to depend on these two areas of scientific and technological endeavor.

At first glance, it may appear that these two topics have relatively little in common. Biotechnology is the study and application of genetic engineering techniques to improve the value of such things as crops, livestock, and pharmaceuticals. It is the adaptation of living systems to produce higher-value-added products and processes. Materials science, on the other hand, involves the study of the properties, design, and wear characteristics of materials, from drugs to alloys, from polymers to ceramics. Prominent areas of investigation with significant commercial stakes include corrosion, pharmaceuticals, and the design of new materials.

Biotechnology and materials science do have something in common: these fields come together through the science of chemistry, the unifying aspect. Chemistry is the molecular science—the study of the composition, structure, and properties of substances and the transformations that they undergo. Form and composition at the molecular level dictate the functioning of all things. How atoms of the 90-some chemical elements that occur in nature link to form molecular structures affects every event in the universe.

Both biotechnology and materials science depend on our ability to manipulate chemical structure. Chemical structure determines the components of all inanimate objects, shapes all forms of life, masterminds our thoughts and actions, dictates health or illness, orders happiness or despair, enriches or impoverishes nations.

The ability to manipulate chemical structure will be critical to our nation's economic vitality in the future. We can better understand this condition if we place it in a global context. A world economy has developed in which every country must compete in global markets to sustain internal growth and economic stability. The developing nations sell low-cost raw materials and commodity products where cheap labor costs provide an advantage. The developed, industrialized nations compete fiercely for markets for their manufactured goods, services, and agricultural products. The issues of tariffs, trade restraints, reciprocal trade agreements, and protectionist legislation are debated daily in the world's capital cities.

For the United States, there are two concerns. Our home markets are being invaded by high-quality, low-cost products from all over the world, and our ability to sell competitively to global customers is eroding. Our rate of productivity gain, although growing at an average annual rate of about 0.3%, is low in comparison to the gains in productivity made by some of our competitors. This situation is partly due to the fact that more nations have attained the "critical mass" in capital, technology, and trained people to begin to catch up to the industrialized nations. However, Japan, an industrialized nation, has an average annual rate of productivity growth that is nine times greater than that of the United States. South Korea's is sixteen times greater.

A study done by the Brookings Institution in 1984 shows that the greatest single contributor to productivity increases is technological innovation (44%). Innovations in technology historically have led to more exports and new jobs. From an understanding of the issues of low-cost natural resources in their countries of origin and the realities of the differential in labor costs between developed and developing countries, the consensus has arisen that technology-based products and services are our hope for remaining a leader and major player in world markets in the next few decades.

Even in high technology, our track record over the past five or six years is not encouraging. In 1980, the high-tech industries produced a trade surplus of $27 billion. The surplus dropped to $4 billion in 1985, and the estimated 1986 figures show a significant deficit.

What then, should we do? A special report in the *Wall Street Journal* of November 10, 1986, entitled "Frontiers of Science—Changing the Ways You'll Live and Work" spells out the prescription:

> Major advances in materials science, genetic engineering, catalytic science, communications,

computers, and artificial intelligence are providing a foundation for changing the methods and even the kind of work people will do in twenty years.... Advanced materials, specialty polymers, [and] ceramics are the absolute core to advanced technologies of the future.... Genetic engineering promises faster plant development and even larger food surpluses .... Scientists are increasingly setting out to make new materials and are succeeding in that task.

The challenge for chemists, and for the nation's policy makers too, is to recognize that chemistry, the molecular science, is at the heart of this new technological thrust. The chemical data bases that chemists built so carefully over the years are the foundation for the molecular design programs now used so aggressively in the development of new drugs, high-performance materials, specialty chemicals, and biotechnology products. These are especially exciting areas for research and development.

This book discusses advances in the new recombinant DNA technology and materials produced by high-technology. Working hypotheses for biological functions can be tested through the deliberate synthesis of tailored molecules: natural product analogs, chemotherapeutical agents, proteins deliberately altered to provide new functions, and genetic inserts. Biochemists are revealing the basic processes of life. The "new" biotechnology will have a significant influence on the nation's economy. It is estimated that by the year 2000, $15–20 billion in new products will be created through recombinant DNA techniques.

Chemical principles and modern experimental techniques now permit systematic chemical strategies for the design of novel materials. These include refractory materials, ceramics, glasses, polymers, and alloys. In the future, we can expect to see entirely new structural materials, liquids with orientational regularity, organic and ionic conductors, self-organizing solids, acentric and refractory materials. Chemistry will be essential to the design of molecular-scale memory and electrical circuit devices, the most dramatic frontier of materials science.

A large fraction of the approximately $175 billion in annual manufacturing shipments by the U.S. chemical and allied products industries, which employ more than one million people, can be attributed to materials science.

The race is on in high technology. Advanced materials, biotechnology, specialty chemicals, and computer-assisted chemistry and biology are essential for the continued vitality of American industry and consequently, for American science.

I hope this book provides you with a glimpse of the exciting chemistry being developed and the economic promise of the fields of biotechnology and advanced materials. The long-term implications are great, both for the United States and for the world at large.

MARY L. GOOD
Des Plaines, IL 60017

Introduction

*B*iotechnology and materials science share a number of attributes. They are both interdisciplinary in nature, with chemistry a component. Aided by new instrumentation and computer modeling, chemists and chemical engineers will have increased roles in these fields in the future. Both biotechnology and materials science are already yielding results of practical importance that will grow greatly in the decade ahead. Progress in each will determine in large measure the relative competitiveness of nations and the health and well-being of their citizens.

In this volume, the authors provide easily comprehensible histories of developments that made possible present-day biotechnology. They detail some of the recent achievements and outline areas where important advances may be expected.

The two major areas that biotechnology will affect are likely to be medicine and agriculture. Already more than 230 diagnostic aids based on monoclonal antibodies have been approved by the Food and Drug Administration. Among them are a home test for pregnancy and a test for AIDS. A number of pharmaceuticals, including insulin, α-interferon (effective against some forms of cancer), and tissue plasminogen activator (an enzyme capable of dissolving clots associated with heart attacks) also have been approved. About forty new pharmaceuticals derived from recombinant DNA technology are under development, and many are undergoing extensive clinical tests. These proteins must be of exquisite purity to avoid unwanted pyrogenic reactions. When approved for use and applied, the new products will have a profound, even revolutionary, effect on clinical medicine.

In the race to develop protein pharmaceuticals and other products of biotechnology, small venture companies led the way. In recent years, large chemical and pharmaceutical companies have become active. For example, DuPont devotes one-third of its billion dollar annual research budget to a broad program including pharmaceuticals and biotechnology. Other chemical companies are active, among them Monsanto, Dow, and American Cyanamid. These companies have long been active in agrochemicals, and they are finding major opportunities in the genetic engineering of plants to produce superior variants, including disease- and pesticide-resistant varieties. The new techniques that are now available are more powerful and effective than earlier approaches based on empirical studies that involved the synthesis of a large number of compounds and extensive tests of effectiveness.

In their chapters, both Simmons and Dervan assert that chemistry will play an essential role in biotechnology. Chemists are especially qualified to tackle problems of determining structure and shape of molecules and their interactions. We are entering an era in which chemistry has the opportunity to solve some of the most important problems in biology and biomedical science. Goddard suggests that a combination of theory, data and computer computations and displays will expedite design of new pharmaceuticals. Scientists already can gain chemical insights by watching on a computer display how a drug docks at an active site on an enzyme. They can explore rapidly the effect of changes in the molecule.

Chemists can take pride in the enormous role that their science has had on industrial production of various materials. For example, an estimated one-sixth of the value of all good manufactured in the United States involved heterogeneous catalysis and other catalytic processes. Economy describes how plastics and ultra-high-strength composites formed using polymers are being improved and are finding steadily increasing applications in airplanes and motor vehicles. It has been estimated that by the year 2000, plastics will have replaced half of the steel in various uses. Ceramics and metals are being improved through better control of their compositions and by use of reinforcing fibers composed, for examples, of carbon or silicon carbide. Whitesides suggests that recent advances in high-temperature superconductivity and in the formation of diamond films point toward technologies that may some day play crucial roles in microelectronics.

In efforts to improve materials, the chemist have available a great variety of analytical instruments that yield precise information about the surfaces and interiors of substances. This information, coupled with fast computers and advanced graphics, is likely to lead to new classes of polymers with specific chemical, electrical, or mechanical properties. Chemists may be able to perfect alloy structures and ceramics. It will be possible to optimize further the selectivity and yield of chemical reactions through improved design of catalysts.

PHILIP H. ABELSON
Washington, DC

Biotechnology

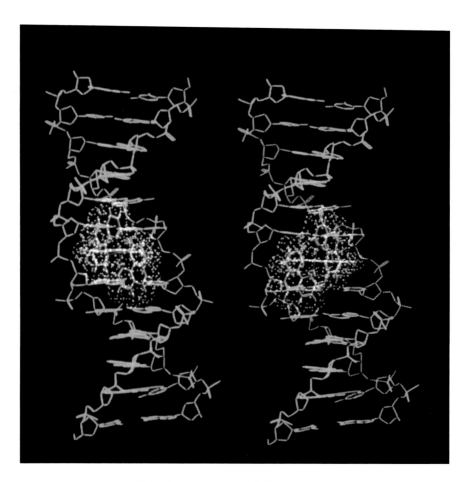

Molecular probes of DNA helical structure.

Chapter 1

What Is Biotechnology?

Jacqueline K. Barton

I have been given the task and opportunity of introducing chapters that describe some of the exciting new chemical advances in the field we call "biotechnology". You may notice that neither the words "chemical" nor "molecular" is incorporated into "biotechnology", but really the heart of what I think is exciting about this area is indeed chemical.

About the Science of Biotechnology

Biotechnology began when chemists learned how to talk about biological processes in molecular terms, in molecular dimensions. An average biological molecule of interest, for example a protein, is from 20 to 50 angstrom units long; that size corresponds to one ten-millionth of an inch. Philosophers have wondered how many angels can dance on the head of a pin. As a scientist, I do not know the answer to that question, but I can tell you that you can put a billion billion average-sized protein molecules on the head of a pin. On that scale, chemists are now designing, searching out, and transforming biological substances to perform quite complicated chemical tasks. So you will find in each of the chapter titles of this section the words "chemical" or "molecular", and indeed the substance of the chemist's approach to this area of science is a molecular description.

Chemistry is not just the name of a scientific discipline. It reflects a particular perspective. Chemists think in terms of molecules: the structures of molecules and their inherent reactivity. *Molecular structure* means the shape of things, the relative arrangement in space of the atoms that are bonded together in the molecule. *Reactivity* means the tendency of a molecule (or a part of a

1473-5/88/0003$06.00/0

molecule) to combine with other molecules to generate new three-dimensional structures. In a chemical reaction, one molecule is transformed into another.

The chemistry described in this section is concerned with biological molecules. These are macromolecules, giant collections of thousands of atoms. Such molecules are therefore orders of magnitude more complex, both in terms of their structure and in terms of their potential reactivity, than the simple molecules chemists have been accustomed to studying.

By virtue of their size and complexity, biological macromolecules can carry out quite complicated chemical transformations. Nature uses large molecules to bring together many different functions in one place and make them work together. That is what consitutes a macro-molecular assembly. In so doing, Nature does quite literally, and actually chemically, convert food into thought.

Can we learn how to harness that chemistry? From my perspective as a chemist, I would say that we are really entering into a chemical revolution, because we are indeed beginning to understand and manipulate biolog-ical molecules in performing these chemical transforma-tions.

About This Book

In the following chapters, some of the scientists who are taking part in this revolution will describe some interest-ing examples of the interplay between chemistry and biology, that which makes up modern biotechnology. Their stories range from the development of chemical methods to read the information contained in human genes to the manipulation of proteins to create new products for industry and medicine. As you read these chapters, you will get a sense of one major direction of chemistry today.

Howard Simmons describes the chemical basis of what is commonly termed biotechnology and goes on to cite several examples of the power of biotechnology in the industrial arena, especially in the area of agricultural research. In the Woody Allen film *Sleeper*, the hero finds himself in a futuristic society, overwhelmed by a ten-foot banana. Agricultural biotechnology promises a much less ridiculous and certainly much more subtle and rewarding future.

The question that Peter Dervan and I are studying in our separate research programs is how one bit of information along a strand of DNA is recognized and accessed, given all the information that is stored along that strand of DNA, all the information that serves as the genetic library of the cell. Can we, with synthetic molecules of our own design, similarly target and selectively and specifically manipulate sites along a strand of DNA? The reading of all the information in that library, a massive task, constitutes the human genome project, which is much in the news of late. It will be akin to transcribing the encyclopedia one letter at a time. Chemistry will provide the tools for this enormous effort, but also for understanding how that information is recognized and used by Nature.

Sometimes a piece of genetic information is missing, altered, or misread. When that happens, disease may occur. Stephen Lippard discusses the principles behind the rational design of new pharmaceutical agents. Every drug has a target, and that target is a large biological molecule. One approach to drug design is to perturb the architecture of that molecular target; the anticancer drug cisplatin may take advantage of this principle. Another approach exploits the specific reactivity of the target. 3'-Azido-3'-deoxythymidine (AZT), the most promising drug to date for AIDS (acquired immune deficiency disease), acts as a chemical monkey wrench in the reactive machinery of an essential part of the AIDS virus. As we begin to describe in molecular terms the architecture and reactivity of biological targets, we will be able to fashion new drugs appropriately, specifically, with the combined rationale of an engineer and a sculptor.

When the genetic library is read properly, that information is used to make proteins. The bits of genetic information encode specific pieces of the protein. E. T. Kaiser recounts how chemists first learned to read both the genetic information and the chemical structure of the protein it produced, the one-dimensional sequence of nucleotides or amino acids. These chemical methods have evolved into tools for the manufacture of specific sequences of DNA, and more recently, for whole protein molecules, some natural, some unnatural. At the heart of chemical methodology is the ability to make molecules.

. . . we are really entering into a chemical revolution, because we are indeed beginning to understand and manipulate biological molecules. . . .

Chemists can now not only replicate the molecules produced by Nature, they can also design and synthesize totally new chemical substances of biological consequence.

Lastly Greg Petsko describes how we can engineer proteins into doing specific tasks of our choosing. We are learning how to convert an antibody, which recognizes a specific foreign substance, into an enzyme, which performs a specific catalytic reaction. Proteins have evolved to carry out specific reactions that are needed in the cell. But the requirements of the cell are different from the requirements of industry or medicine. We are learning how to redesign proteins to carry out tasks useful for the chemical industry, for the food industry, and even for your family doctor.

Nature, after all, *is the best chemist we know.* We are learning how to exploit Nature's chemistry. That's what biotechnology means.

Chapter 2

Biotechnology: A New Marriage of Chemistry and Biology

Howard E. Simmons

This is an exciting time in science. New discoveries are being made with awe-inspiring frequency in every branch of science, and those discoveries are changing our perceptions of ourselves and our universe, not to mention providing the basis for important new technologies. Fields as diverse as materials science, particle physics, and astronomy all are progressing rapidly. Research into high-temperature superconductivity—a scientific revolution happening before our very eyes—promises remarkable benefits to mankind.

How Biotechnology Is Improving Our Lives

Despite the exciting progress occurring in these scientific disciplines, probably today's most significant and profound scientific advances are those occurring in molecular biology. Biotechnology, which seeks to apply this new understanding of the chemistry of life for the good of mankind, has already begun to change our lives. Five pharmaceutical products produced by recombinant DNA techniques are on the market:

- human insulin
- human growth hormone
- α-interferon
- hepatitis B vaccine
- tissue plasminogen activator (TPA)

A number of diagnostic products based on monoclonal antibodies also have reached the market, and many others are on the way. Those already available cover a range as diverse as home test kits to determine pregnancy

1473–5/88/0007$06.00/0

to the highly publicized test to screen blood for anti-bodies to human immunodeficiency virus (HIV), which causes acquired immune deficiency syndrome (AIDS).

Many more pharmaceutical products are on the way. Academic and industrial researchers have targeted nearly every major human disease and already have discovered proteins that appear useful in clinical trials for treating those diseases. A compound called tissue plasminogen activator, for example, eventually will help prevent the deaths of thousands of heart attack victims. In treating cancer, a host of natural proteins—interferons, interleukins, colony-stimulating factors, and tumor necrosis factor—affect immune system function and have shown varying degrees of promise in aiding the body's natural ability to recognize and attack malignant cells. Other products now being tested will benefit people ranging from kidney dialysis patients to hemophiliacs.

Agricultural biotechnology will have an enormous effect on the practice of agriculture. Broadly defined, it encompasses vaccines and treatments for animal diseases, and products that will change the productive performance of farm animals. It involves plants with an array of new characteristics. These include plants with built-in resistance to herbicides, plants that can produce their own insecticides or fix nitrogen, and plants that produce seeds and fruit with improved nutritional or processing properties. Agricultural biotechnology also includes genetically engineered microbes with beneficial traits such as highly specific insecticidal activity or the ability to limit frost damage to sensitive crops like strawberries.

The progress already evident today and predicted for the future comes from the wedding of biology and chemistry.

Biotechnology, however, is more than a collection of individual products. It is a set of tools that has altered in a very basic way the capabilities of researchers trying to understand life. Perhaps no other problem facing humanity better illustrates this change than the disease AIDS. The techniques of biotechnology have been absolutely essential in developing our current understanding of the virus that causes AIDS, and they will be equally essential in developing treatments for the disease and a vaccine against it. As more than one AIDS researcher has noted, if the disease had struck a mere 20 years ago, it probably would have remained a terrifying plague of unknown origin. Instead, researchers have progressed more rapidly in their understanding of AIDS, which is a bewilderingly complex disease, than they have against any other disease in history.

And there is much, much more to come.

Chemistry Underlies Modern Biology

Nonscientists hear the terms "biotechnology" and "molecular biology", and not surprisingly, they often think that biology dominates these remarkable advances. But that isn't so. Chemistry, in fact, is the field that underlies modern biology. The progress already evident today and predicted for the future comes from the wedding of biology and chemistry. Chemistry's essential contribution to this marriage is the ability to unravel biological mechanisms at the molecular level and the ability to manipulate molecules, even ones as large and complex as the macromolecules that dominate biological processes.

Origin of Life

Let's try to make this a bit clearer by looking at the origin of life. Although we do not know exactly how it happened, the beginnings of life on this planet could not have occurred until Nature had created an array of complex organic molecules from the very simple ones that first formed as the Earth cooled. Somehow, those molecules assembled into units that could reproduce themselves—the first living organisms. After that, we are on surer ground, because we understand how evolution operated on those organisms. Evolution gave rise to a very wide variety of complex biological systems, the most elegant and important of which is the system that uses deoxyribonucleic acid (DNA) and ribonucleic acid (RNA) to carry the information that guides the formation of each new generation of all organisms from the simplest bacteria and slime molds to majestic redwoods to humans themselves. Throughout about three billion years of evolution, Nature has been altering DNA—a process we call *mutation*—and creating new species through natural selection.

The Genetic Process

Figure 1 depicts schematically how this genetic process works. The DNA molecule located in the nucleus of a cell carries in its structure the instructions that enable each living cell to function in its proper way. The DNA molecule serves as a template to guide the formation of smaller molecules of messenger RNA (mRNA) in a process called transcription. These mRNA molecules then

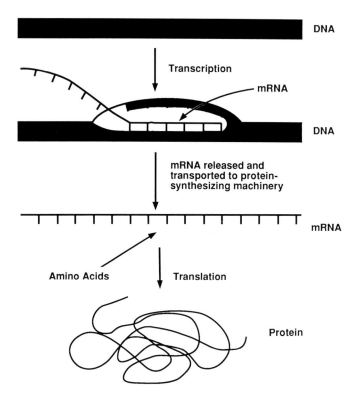

Figure 1. Mechanism of gene expression.

leave the cell nucleus and act as a second template for the assembly of amino acids into proteins in a process called translation. Proteins, some of them modified with sugar molecules, carry out many functions in organisms. One of their most important roles is as enzymes, or biological catalysts, which serve as key controllers of a wide range of biological systems.

Cross Breeding

Sometime before the beginning of recorded history, one of mankind's most important scientific revolutions began: The domestication of plants and animals by cross breeding. That process—which is, in fact, the first example of humans deliberately altering DNA—led to many new living species. Some possessed desirable characteristics; witness, for example, our familiar lines of cattle, sheep, dogs, ducks, fruits, grains, and many others.

Although very effective, this process of breeding was empirical—in other words, a trial-and-error process—and slow. Until fairly recently, there was obviously no understanding of the breeding process in what we would call a scientific sense. Gradually, humans developed a scientific understanding of these biological processes at the whole organism level. Gregor Mendel's discoveries in the genetics of garden peas are an elegant example. Mendel observed that specific traits such as tall or short pea plants or smooth or wrinkled peas segregated independently and in a statistically predictable way. From these observations, he derived the rules of classical genetics.

Charles Darwin provides another example. Although observations of wild living creatures by Darwin on his famous voyage to South America form much of the argument put forth in *The Origin of Species,* the first chapter of this seminal book deals with variation under domestication. Indeed, Darwin's comprehensive understanding of breeding domestic animals, particularly pigeons, forms a cornerstone of his theory of evolution.

Convergence of Chemistry and Biology

As chemistry developed an understanding of molecules and their behavior, these biological processes began to be described at a more basic level. Both chemists and biologists over the past 200 years have contributed tremendously to this understanding. However, all this science—both biological and chemical—was primarily descriptive when it came to living things. Scientists began to understand how biological systems worked, but the knowledge was insufficient to allow them to alter the fundamental processes going on in the systems.

Quite separately, other chemists over the past two centuries developed exquisite skills in controlling the reactions of myriad molecules not involved in biological processes. Although some of these molecules were quite complex, they were, in general, much simpler than a typical protein or nucleic acid molecule. The vastly productive field of synthetic chemistry has yielded dyes, plastics, synthetic fibers, agrochemicals, and pharmaceuticals. It is a truly endless list of useful products.

The revolution in molecular biology has resulted from the convergence of these two streams of science—the inherently descriptive biology and biochemistry that has provided a considerable, though far from complete, understanding of how living systems operate at the molecular level, and the synthetic and mechanistic chemistry that allows the manipulation of those molecules in useful ways. The term *biotechnology* was coined to describe the application of this merging of the descriptive and the synthetic sciences.

Basis of Biotechnology

Biotechnology stems from the relatively recent elucidation of the role of DNA as the storehouse of the genetic code and the determination of its molecular structure. Figure 2 provides two versions of that molecular structure. It is important to realize that scientists today have a very accurate understanding of the basic structure of DNA and the many subtle permutations of that structure.

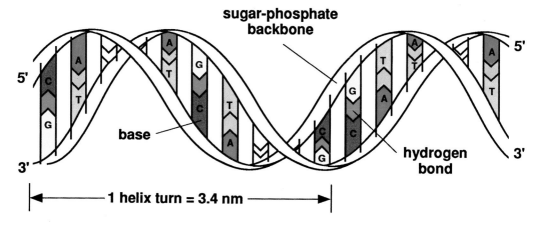

Figure 2A. The DNA double helix.

Once researchers understood the molecular details of DNA and gained a clear sense of how it functioned, the opportunity to use the synthetic and mechanistic skills of the chemist to modify these molecules became a reality. Of course, biologists, molecular biologists, biochemists, and chemists of many different stripes are still working very hard to understand more about the details of how DNA is expressed, regulated, and perhaps most importantly, how that process sometimes goes awry in various disease states, but it is accurate to state that the time for the fruitful marriage of chemistry and biology has arrived.

An extremely important demonstration of the power of controlled modification of the DNA molecule came with the discovery of the laboratory process referred to as *recombinant DNA technology*. This process uses natural enzymes—protein molecules with names like "ligases" and "restriction enzymes" that clip DNA and sew it back together again—to carry out controlled chemical reactions on DNA molecules. That is the same thing as saying that scientists can carry out controlled chemical reactions on genes, the elements that carry the blueprints that define every living creature. These altered, or recombinant, DNA molecules then carry a new genetic message to the species into which they are placed. These foreign genes may be expressed in their new host so that, in effect, one species can produce proteins normally formed by another species.

At its simplest level, such a transformation allows, for example, the common bacterium *Escherichia coli* to produce human insulin. Most early research in recombinant DNA involved just such manipulations of cells. At a more complex level, desirable proteins can be expressed in more complicated cells such as cultures of mammalian fibroblasts. At yet a more complex level only now being attained, a trait controlled by a single gene can be expressed in an entire organism like a tomato plant or a cow.

Figure 2B. Molecular structure of DNA.

Significance of Biotechnology for Industry

What significance does all this have for industry? Obviously, the pharmaceutical industry—made up of large, established drug companies and the numerous, entrepreneurial biotechnology start-up firms—have embraced biotechnology wholeheartedly. It would be wrong, however, to think that the chemical industry is any less excited about the prospects of biotechnology.

People who work in the chemical industry are proud of the many important contributions to society they have been able to make by application of chemical knowledge. Du Pont, for example, at the beginning of this century, turned the chemical skills it had developed in the explosives field to such new areas as synthetic dyes, paints, fibers, and films. Out of such practical application of chemical knowledge has grown the American chemical industry, one that is still earning a substantial positive trade balance for the United States.

The chemical industry now sees a new growth opportunity in this wedding of chemistry and biology— the new field of biotechnology. To illustrate its impact: One third of Du Pont's billion-dollar annual research budget now goes into biotechnology, broadly defined, and this level of research support is indicative of the trend throughout the chemical industry.

Many chemical companies perceive this as a logical development, simply the next outlet for their chemical skills. Simply put, the biotechnology revolution is every bit as much chemical as biological.

Du Pont's experience is illustrative. The company's Central Research & Development Department has the responsibility for the long-range, basic science aspects of Du Pont's R&D effort. In support of the company's expansion into the life sciences, this department now devotes about half of its research to this area. A key point is that half of the scientists working on these research programs are chemists. Biotechnology is truly a partnership of these two major fields of science.

How broadly is biotechnology affecting the chemical industry? Dow, Monsanto, and American Cyanamid, to mention only a few, are following courses similar to that taken by Du Pont. Monsanto, for example, has invested more than $150 million in its Life Sciences Research Center, a facility wholly devoted to biotechnology research. Even companies not strictly thought of as chemical companies, but having a high level of chemical expertise, are pursuing biotechnological goals. Eastman Kodak is a good example.

Du Pont views biotechnology as having two primary applications, agriculture and the health sciences. Many chemical companies have been in the agrochemical business for a long time, so the application of biotechnology to agriculture probably falls into the category of evolution rather than revolution. Nevertheless, the changes biotechnology will bring to agriculture will be dramatic.

Designing Agrochemicals

For many years, the process of finding new agrochemicals was largely empirical—sophisticated trial and error, but trial and error nonetheless. Once chemists discovered a chemical structure that seemed effective, they would synthesize a great many closely related compounds to find the most effective variation. What has changed? Primarily, the ability to use our new molecular knowledge of biological systems to *design* chemicals to accomplish a desired result.

Sulfonylurea Herbicides

The transition can be illustrated by a class of Du Pont herbicides called sulfonylureas. Du Pont chemists discovered this class of chemicals by empirical screening. The sulfonylureas kill weeds at amazingly low application rates—literally fractions of an ounce per acre—and they have very low animal toxicity. But why? Biological research provides the answers. As shown schematically in Figure 3, the sulfonylureas tie up an enzyme called acetolactase synthase, which is often called ALS for short. This enzyme catalyzes the first step in the biosynthesis of the branched-chain amino acids—valine, leucine, and isoleucine. Plants must be able to synthesize these amino acids from carbohydrates produced by photosynthesis, or they die. Because a catalyst is a material that causes a chemical reaction to occur but it not used up itself in the reaction, only very small amounts of ALS are present in the plant. Therefore, only a very small amount of the chemical used to tie up the catalyst—in this case the sulfonylurea—need be applied. Hence the low application rates.

The low animal toxicity of the sulfonylurea herbicides also can be explained through biochemistry. It turns out that animals do not have ALS. Humans obtain the branched-chain amino acids, and several other so-called essential amino acids, in their diet. Because ALS is not present in animals, compounds whose toxicity depends on blocking the activity of ALS are not particularly toxic to animals.

Du Pont's understanding that the function of the herbicide molecule is to block the catalytic activity of ALS has helped the company synthesize and patent many variations on the original sulfonylurea herbicide. Generally, in the story of a herbicide, that is where it ends. The herbicide is effective, it is safe, and farmers use it. But there is still a challenge. Wheat and other cereal grains are resistant to the sulfonylureas, but corn is not. Again, a look at the biological mechanism of sulfonylurea

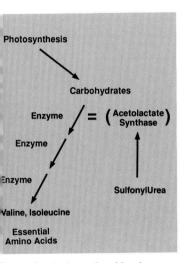

Figure 3. Action of sulfonylurea herbicides.

activity at the molecular level provides an explanation. Wheat produces another enzyme that metabolizes—or chews up—the sulfonylurea molecules and prevents them from tying up the ALS molecules.

This discovery suggests that it might be possible, by using the techniques of biotechnology, to alter corn to make it resistant to sulfonylurea herbicides as well. One strategy that suggests itself is to isolate the gene in wheat that encodes the enzyme that metabolizes ALS and introduce it into corn plants. There is another strategy, however, that Du Pont scientists also have pursued. Bacteria also synthesize the branched-chain amino acids by using a biosynthetic pathway that involves ALS. Some strains of, for example, *Escherichia coli*, however, are resistant to the toxicity of the sulfonylureas. Scientists at Du Pont and, independently, at American Cyanamid, which is also working on sulfonylurea resistance in plants, isolated the gene coding for the resistant ALS and determined its sequence. They found that a single base pair was different in this DNA molecule from the plant DNA. As a result, the resistant *E. coli* produced an ALS enzyme containing, in a specific position, a threonine amino acid residue instead of an alanine amino acid residue. Replacing that one amino acid produces an ALS molecule that does not interact with the sulfonylureas in the same way that plant ALS molecules do. So another means of producing corn that is resistant to sulfonylureas would be to introduce the gene for the resistant ALS molecule into corn plants.

Glyphosate Herbicides

A similar story can be told for the glyphosate herbicides produced by Monsanto. Glyphosate works by inhibiting another enzyme unique to plants, 5-enol pyruvyl shikimate-3-synthase (EPSP synthase), which catalyzes a critical step in the synthesis of aromatic amino acids like phenylalanine, tyrosine, and tryptophan. Again, animals do not synthesize aromatic amino acids, so glyphosate has very low animal toxicity. But glyphosate is almost entirely nonselective—it kills all green plants. Researchers at Monsanto have developed a method for enhancing a plant's ability to produce EPSP synthase 20 times over the normal level. Such plants exhibit resistance to normal levels of glyphosate application. Researchers at Calgene, a plant biotechnology company in California, have taken the alternative route to producing plants with glyphosate resistance. They have discovered a EPSP synthase in bacteria that is naturally resistant to glyphosate.

Some biotechnology critics ask why companies like Du Pont bother producing herbicide-resistant crops in an era of farm surpluses. These crops will allow the adoption of low-cost, broad-spectrum, environmentally safer herbicides. Such practices are important, especially when farmers are producing large surpluses of food, to reduce the cost to the farmer, and not incidentally, to improve environmental quality. There is another reason, of course, an economic one—such herbicide-resistant crops will broaden the use of proprietary products.

Introducing Genes into Plants

These are very difficult experiments to carry out, and in fact, all of the technology for the work just described has yet to be worked out. Individual genes of interest usually can be found and isolated. Getting them into a plant, however, is not so easy. Scientists have discovered techniques for putting genes into some plants, principally broadleafs like tobacco, tomato, and soybean plants. This method involves a naturally occurring bacterium called *Agrobacterium tumefaciens,* which is a natural genetic engineer. In nature, this bacterium is able to insert a piece of its DNA into the chromosome of a plant, which causes what is called crown gall disease. Crown galls are tumorlike growths on plants. Scientists utilize *A. tumefaciens'* genetic machinery to move genes of their choice into plants. However, the bacterium's machinery does not work on the cereal plants—corn, wheat, and rice among many others—which are so important to agriculture the world over. Other elegant techniques that have been developed to introduce genetic material into plants include microinjection and, as unbelievable as it may sound, a laboratory shotgun that literally blasts tungsten beads coated with DNA into plant cells. Yet another problem facing researchers is reproducing a whole plant from the single cells that are often the target of these gene introduction methods. Progress is being made, however, by academic and industrial researchers on all these problems.

Applications in the Health Sciences

Equally great opportunities exist for chemical companies to use the same kind of collaboration among chemists and biologists in health sciences. As pointed out, research on AIDS is an area where biotechnology has played an important role. At Du Pont, for example, scientists made a basic contribution in determining the structure of certain parts of the HIV genome. On the applied side, Du Pont has developed diagnostic systems for AIDS and is continuing to work on improvements in these systems. These are not insignificant business opportunities. Estimates place the total market for the HIV-antibody test, for example, at $75 million in 1986.

Yet another illustration of basic research at Du Pont is shown schematically in Figure 4. On the left is a representation of a gene, one segment of an entire DNA molecule, which directs the production of an important natural enzyme called β-lactamase. This enzyme, which metabolizes penicillin, is what gives some bacteria resistance to that antibiotic. Three DNA bases are highlighted in the chain—designated A, B, and C—and these code for a specific amino acid. Using recombinant DNA techniques, Du Pont scientists removed that ABC unit, chemically replaced one of the bases—represented by replacing the B with an X—and reinserted it into the gene. The new enzyme, called β-thiolactamase, has one amino acid out of 128 in its molecular chain different from the natural enzyme. In fact, the net change in the whole 2000-atom enzyme is the replacement of a single oxygen atom by a sulfur atom. That, however, subtly

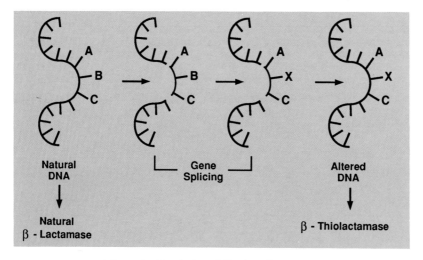

Figure 4. Chemical modification of genes.

alters the enzyme's biological behavior. This example demonstrates the tremendous potential for using chemistry in a very sophisticated way to tailor biological systems for useful purposes.

Summary

This, then, is why chemists and chemical companies are so excited about biotechnology. Effective utilization of biology's emerging knowledge of the molecular mechanisms of life will depend heavily on chemistry's capabilities in designing and altering molecules. The potential benefits to society are truly immense. Important steps toward these ends already are taking place in many chemical laboratories, and some have moved out into the practical world. If this nation supports those who are bringing these two sciences together, future generations will benefit handsomely.

Chapter 3

...tic Tools for Molecular Biology

Peter B. Dervan

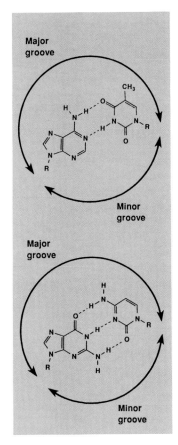

Major groove

Minor groove

Major groove

Minor groove

*C*hemistry has made tremendous advances over the past four decades in the broad fields of synthesis and understanding chemical reactivity. In that same time span, a series of revolutionary events occurred in biology. First came the discovery of the double helical structure of DNA in the 1950s by Watson and Crick. This discovery allowed the elucidation of the mechanisms of DNA replication—how DNA makes copies of itself—and DNA transcription and translation—the processes that allow the genetic code to be read and translated into proteins. In the 1970s, the techniques that permit DNA to be cut and spliced in controlled and well-defined ways were invented and the technology of recombinant DNA was born.

Chemistry plays a pivotal role in this biological revolution because biological events involve molecules and molecular interactions. No scientists are better qualified to tackle the problem of determining the structure and the shape of molecules than chemists. Our society is entering an era in which chemists have the opportunity to solve some of the most important problems in biology and biomedical science.

Each strand of DNA consists of a linear polymer of nucleotides, which are phosphorylated versions of four different bases: adenine (A), guanine (G), cytosine (C), and thymidine (T). Physical chemical concepts readily explain what holds these nucleotides to complimentary nucleotides on the second strand of the DNA molecules. Hydrogen bonding dictates that the base adenine matches up with the base thymidine and the base guanine matches up with the base cytosine (*see* structures). These basic principles of chemistry make it possible for scientists to explain, in a very fundamental way, how life works.

The quantity of DNA that encodes the entire human genome consists of about three billion such base pairs.

1473–5/88/0021 $06.00/0   © 1988 American Chemical Society

That is three billion chemical bits of information. These chromosomes—humans have 23 pairs—are the information repository, or encyclopedia, for the construction of a human being. Molecular biologists estimate that there are 100,000 to 300,000 segments of information or genes contained on these 23 pairs of chromosomes.

Each gene, typically, is the blueprint for one protein. The three-dimensional shape of proteins gives our bodies form and function and allows the proteins to carry out the complex chemistry of life. Clearly, then, understanding the structure of human chromosomes and genes is one of the first steps toward understanding the fundamental machinery of human existence. We already understand a great deal about those structures, but there is much more to learn. These are enormous molecules, and the subtle and complex interactions among them determine the difference, for example, between health and disease.

During the past few years, there has been considerable discussion about whether scientists should begin the task of first mapping and then sequencing the human genome. This is a task of staggering dimensions. *Physical mapping* means using restriction endonucleases to cut up chromosomes and then determine the order of the resulting fragments. *Genetic mapping,* a somewhat less daunting task that already is being undertaken, means locating on chromosomes about 400 known genetic markers. *Sequencing* is the analysis of the order of each of the three billion base pairs that make up the human genome.

Sequencing the Human Genome

First, researchers will have to separate the 23 human chromosomes and obtain a significant quantity of each one. Once the chromosomes have been sorted, one strategy that has been proposed is to break each chromosome into fragments, each fragment containing about 40,000 base pairs. Each chromosome would yield about 2500 of these fragments, and their order on the chromosomes would then have to be determined. These fragments, in turn, would have to be broken down into fragments of a suitable size for direct sequencing, about 1000 base pairs long. Each chromosome would yield about 100,000 such fragments. Simple arithmetic shows that, at a rate of about one million bases a day, sequencing the human genome will take 10 years to complete.

The potential applications of possessing the sequence of the entire human genome are enormous. These applications include such things as the ability to gain a basic understanding of why certain individuals are susceptible to genetic disorders like cancer, heart disease, and mental illness.

At Caltech, biology professor Leroy Hood and his co-workers have worked on developing the chemistry and instrumentation for sequencing DNA and proteins. Hood would be the first to acknowledge that current automated DNA sequencing is still a relatively immature technology and that it will require significant further development in both the underlying chemistry and the instrumentation before it is ready to take on the task of sequencing the human genome. That work is being done both at Caltech in Hood's group and more recently by a group at DuPont. The step toward sequencing the human genome should be the automation of the entire sequencing procedure, not just the downstream portion of sequencing nucleotides, an effort that may take five to ten years. Within that time, nucleotide sequencing should have improved in speed by a factor of about 100 and improved significantly in accuracy. It should then be ready to be applied to the human genome.

The potential applications of possessing the sequence of the entire human genome are enormous.

During the 15 to 20 years it will take to map and sequence the human encyclopedia, it is probable that the fields of molecular biology and biological chemistry will make new strides into understanding the mechanisms that control gene expression. Gene expression is the product of an extensive array of interactions among DNA, RNA, and proteins. Chemists want to understand the physical and chemical principles that govern this process. One aspect of this research will be to isolate and determine the structures of protein–DNA assemblies that regulate how genes are expressed. It is conceivable, therefore, that concurrent with the determination of the sequence of human DNA, the "software" of the human genome also will be understood, and synthetic chemists will be able to engineer at the molecular level novel synthetic materials that read unique DNA sequences.

For instance, the natural restriction endonucleases that have been used so powerfully by molecular biologists in recombinant DNA techniques can recognize a specific DNA sequence, four to eight base pairs in size, and cut the DNA molecule at that site. A consequence of a four-letter alphabet for double helical DNA is that each binding site size yields a specific number of unique sites. There are, for example, 136 unique binding sites consisting of four base pairs and 2080 unique sites consisting of six base pairs. In terms of physical mapping of the genome, this number of unique binding sites limits what might be called the resolution of the map.

Improving Restriction Endonucleases

Natural restriction endonucleases recognize and bind to sequences in DNA from four to eight base pairs long and then cut the DNA at that site. A question synthetic chemists in my laboratory have addressed is whether we can improve on the specificity of natural restriction endonucleases. In other words, is it possible to modify natural DNA-binding molecules or synthesize entirely new one that will recognize unique sequences of 12 to 15 base pairs? Again, because DNA uses a four-letter alphabet, there are 10 million to 100 million such unique DNA sequences, and hence, one could potentially create that many molecules. To be specific at the level of one unique gene in large chromosomal DNA, molecules need to recognize DNA sequences made up of 15 base pairs.

This is a chemistry problem in molecular recognition. Despite all the successes of chemistry in the past 30–40 years, the field of molecular recognition in organic chemistry—how macromolecules fold and fit together—is still in its infancy. The way to solve this problem is not by serendipity or by focusing on one particular sequence, but rather to work toward a general solution. That is, to elaborate a set of chemical principles that govern DNA recognition without trying to anticipate the direct application of the DNA-binding molecule in each case.

However, the eventual applications are important. Once scientists have in hand the physical map of the human genome, and for example, know where the gene that causes a particular genetic disease is located, the question is, What are they going to do about it? If the principles of gene expression and regulation are known, then most likely chemists will be able to build artificial repressors for specific DNA sequences to control certain disease states.

Because the understanding of the chemistry of molecular complexation is still primitive, the initial efforts in this research borrow heavily from nature. Numerous natural compounds bind to DNA. The restriction endonucleases, as we have seen, bind and cleave DNA. A variety of small molecules, many of them drugs having antibiotic, antiviral, or antitumor activity, also bind to DNA and some also cleave it. Their pharmacological activity results, presumably, from their ability to interfere with some aspect of DNA's function or repair. The goal is to take such small molecules that bind DNA in a modest fashion and improve them to new and novel specificities.

To screen a large number of potential binding sites on DNA, we couple the binding event to the analytical power of gel electrophoresis. This step is accomplished by attaching to the DNA-binding molecule another moiety that cleaves DNA. Thus, we create molecules with two functions: the ability to bind to specific DNA sequences and then the ability to cut the DNA at points adjacent to that sequence. These bifunctional molecules form the basis of a technique called *affinity cleaving,* that is, the binding event on DNA is converted to a sequencing event (*see* Figure 1). Identification of the preferred binding sites of our designed synthetic molecules is a first step toward addressing the underlying principles of recognition of DNA at the molecular level.

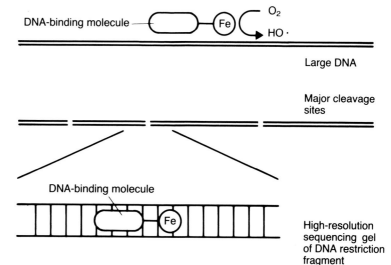

Figure 1. Affinity cleaving technique combines DNA-binding molecule with a group that cuts DNA backbone.

Natural products we have investigated are netropsin and distamycin, di- and tripeptides that are known to bind in the minor groove of right-handed DNA rich in adenine and thymidine bases. The specific interactions involved in this binding process have been determined from the high-resolution crystal structures of a netropsin- and distamycin–oligonucleotide complexes (*see* Figure 2). The question is whether we can improve and alter the specificity of this natural product by building analogs of the molecule that would bind larger DNA fragments. One simple strategy is to build longer versions of the natural product. The longer versions are peptides that consist of four to nine amino acids designed after the structure of the tripeptide. These longer molecules based on distamycin do, in fact, possess a higher specificity for DNA approaching that of restriction endonucleases. So we are moving in the right direction.

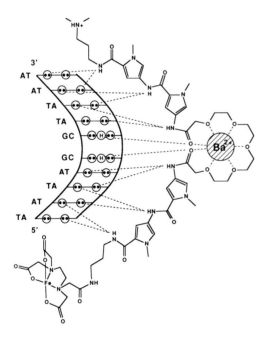

Figure 2. Two netropsin analogs connected by a tetraethylene glycol tether bind a 10 base-pair sequence in the presence of barium metal cation by forming hydrogen bonds (dotted lines) with lone pair electrons ●. The EDTA–Fe group is highlighted.

Molecular Engineering

Another generation of these synthetic molecules will be derived from our understanding of how proteins bind DNA. It might be possible to design synthetic hybrid protein molecules with two structural and functional domains, that is, create synthetic peptides derived from two different proteins. Researchers in our group have combined 52 amino acids derived from the natural enzyme, Hin-recombinase, which contains 190 amino acids, with three amino acids that constitute the copper binding site in serum albumin. This hybrid synthetic protein possesses two separate structural and functional domains. The 52-amino acid fragment provides the DNA binding specificity of the Hin protein and the three-amino acid fragment binds copper and provides oxidative cleavage of DNA in the presence of hydrogen peroxide and ascorbate.

We have also constructed oligonucleotides to which is attached an EDTA–iron DNA-cleaving function. We have shown that certain such oligonucleotides (15 bases long) will form a triple helix in the major groove of double helical DNA (Figure 3) 11 to 15 base pairs long. This finding makes possible a general solution to the problem of sequence-specific DNA recognition. Within certain constraints, we can now build molecules that can recognize unique specific stretches of DNA 15 base pairs long, which is the level of specificity we need to recognize individual genes in the human genome.

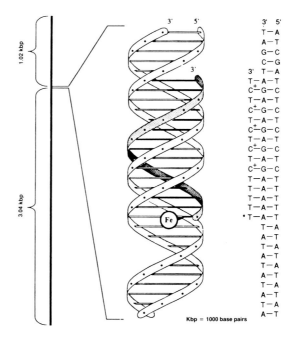

Figure 3. Oligonucleotide recognizes 15-base-pair sequence in 4000-base-pair DNA fragment by forming a triple helix with double-stranded DNA.

Catalytic Antibodies

Turning from the problem of sequence-specific DNA recognition, let me focus briefly on protein recognition and monoclonal antibodies, another important leg of biotechnology. Antibodies are the protein molecules produced by an organism's immune system that bind to foreign molecules in the body and tag them as foreign so that they can be eliminated. In immunological terms, the foreign molecule is an antigen. In an intact organism, a range of subtly different antibodies can be generated for any given antigen; there are, in other words, a number of ways to solve the problem of binding a given antigen. This condition is called a polyclonal response. Antibody-producing B lymphocytes can be fused with a cultured myeloma cell to produce a hybridoma that secretes a specific monoclonal antibody (1).

Recently, independent research performed at the Research Institute of Scripps Clinic (2) and at the University of California, Berkeley (3), shows that, through careful design of an antigen, one can produce an antibody that catalyzes a chemical reaction. Schultz at Berkeley and Lerner and Tramontano at Scripps produced antibodies that catalyze, with a high degree of specificity, hydrolysis of esters and carbonates. The antigenic molecules used to elicit these antibodies are phosphonates, and they mimic the transition state of this hydrolysis reaction.

These researchers are exploring ways to adapt this idea to devise molecules that will cleave protein molecules with great specificity. One direction the researchers are pursuing to introduce catalytic activity into antibodies is production of what might be called *semisynthetic catalytic antibodies*. The idea is to produce an antibody that binds a specific molecule, and chemically to attach catalytic groups such as a metal ion to the antibody to carry out a reaction such as hydrolysis of the peptide bond that links amino acids together in a protein. In such a molecule, the antibody provides the binding specificity and synthetic catalyst provides the needed chemistry.

The Future

In short, when the chemical composition of the hardware and software of the human cell is described at the molecular level, the potential will exist to synthesize molecules to control disease in a very precise way. Over the next 20 years, chemists will participate with biologists in solving these extremely important and challenging problems. The solutions will have tremendous practical implications for humanity.

Acknowledgments

The author is grateful to the National Institutes of Health, the American Cancer Society, the DARPA University Initiative Research Program, Allied Signal Corporation, Merck Sharp & Dohme, Research Laboratories, Burroughs-Wellcome Company and the Ralph M. Parsons Foundation for generous support. In addition, stimulating and helpful discussions with Dr. Ralph Hirschmann are gratefully acknowledged.

References

1. Milstein, C.; Köhler, G. *Nature,* **1975,** *257,* 495.
2. Tramontano, A.; Janda, K. D.; Lerner, R. A. *Science,* **1986,** *234,* 1566.
3. Pollack, S. J.; Jacobs, J. W.; Schultz, P. G. *Science,* **1986,** *234,* 1570.

Chapter 4

Molecular Basis of Drug Design

Stephen J. Lippard

*T*here is no clearer illustration of the beneficial influence of natural science, especially chemistry, on society than the remarkable increase in life expectancy, from about 40 years to more than 70 years, in the period following the introduction of antibiotics and other drugs to treat infectious disease. It is interesting to contemplate the loss to humanity of the creative contributions that might have issued forth had tuberculosis not taken Shelley at age 30, Keats at 36, Heine at 59, Emily Brontë at 22, Ann Brontë at 29, D. H. Lawrence at 45, and George Orwell at 47; or had syphilis not claimed Schumann at 39, Chopin at 39, Nietzsche at 56, and Gauguin at 55. Syphilis was eventually conquered by the "magic bullet" of Paul Ehrlich, who tried 605 arsenic-containing compounds before discovering number 606 to be effective (*1*).

Today the odds against finding a single effective pharmaceutical by such an empirical approach have lengthened considerably, to about 7000 to 1. The average cost to the drug companies of discovering and bringing a single drug to market now approaches $100 million, because once the chemistry and pharmacology have been completed, promising compounds must undergo extensive toxicological studies and clinical trials.

Specificity: Key to Drug Development

Modern chemistry, biochemistry, and molecular biology are moving rapidly toward the day when scientists will no longer have to rely upon trial and error to prepare new molecules for the treatment of human disease. From the knowledge of the molecular structures of biological targets, it is becoming possible to design and synthesize compounds as effective chemotherapeutic agents. What we can do now is very sophisticated compared to Ehrlich's time. In particular, we have begun to understand the fundamental relationships linking the chemotherapeutic potential of a molecule with its ability to recognize and react chemically with a specificity that is dictated by the three-dimensional architecture of the target molecule in the cell. This concept of *specificity* is key to developing drugs that are toxic to diseased tissue and relatively nontoxic to normal cells, or that interfere with the biochemistry of an infectious viral agent without disrupting normal cellular functions.

Cisplatin, a Basic Research Discovery

Current knowledge about the molecular mechanism of action of *cis*-diamminedichloroplatinum(II) (*cis*-DDP), or cisplatin, which is today a leading drug for the management of testicular, ovarian, and head and neck cancers, illustrates this point. Although this simple inorganic compound has been known for more than 140 years, its chemotherapeutic potential was only recognized in the late 1960s following a serendipitous discovery made by Barnett Rosenberg, who was at that time a biophysicist working at Michigan State University. Rosenberg and his co-workers were studying the effects of electric fields on bacterial cells growing in culture. Under the conditions of their experiment, the researchers observed that the cells stopped dividing and formed long filaments (2). Subsequently, the researchers found that this phenomenon had nothing at all to do with the electric fields. Instead, it resulted from the presence of certain platinum compounds in the culture fluid in which the bacteria were growing. These platinum compounds formed by a reaction between ammonium chloride present in the buffer and the platinum metal used in the electrodes that applied the electric field. Addition of one such compound, cisplatin, to the cells in the absence of an electric field reproduced the filamentous growth.

Because it had been known that compounds that induce filamentation in bacteria often had antitumor activity, cisplatin was then tested against various tumors in experimental animals and found to be extremely effective. Human clinical trials commenced in the early 1970s, and the drug received approval from the Food & Drug Administration in 1979.

This story provides a striking example of how

This story provides a striking example of how fundamental scientific inquiry can lead ultimately to the welfare of mankind.

fundamental scientific inquiry can lead ultimately to the welfare of mankind. Nonscientists sometimes ridicule experiments because what is being studied seems so arcane. One can imagine Rosenberg's research coming under such fire. Who cares, such a critic might ask, about how electric fields affect bacteria? It is simply not possible to know, a priori, when research into such questions will turn up a piece of information with enormous potential benefits to society.

Cisplatin Isomers

One intriguing aspect of the biological activity of cisplatin is that its geometric isomer, *trans*-diamminedichloroplatinum(II) (*trans*-DDP) is inactive (*see* structures). They are very similar molecules, but their properties in tumor cells are very different.

Recent studies in our laboratory suggest an explanation for the different chemotherapeutic properties of the two isomers. This explanation is based on an understanding at the molecular level of their interactions with DNA, the proposed target of these chemicals in the cancer cell (*3*). As is so often the case in chemistry, a precise knowledge of the geometry of an interaction between two molecules is required to understand a mechanism.

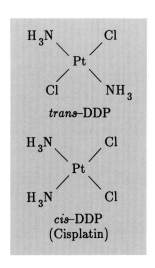

trans-DDP

cis-DDP
(Cisplatin)

Cisplatin Binds to DNA

Cisplatin is administered in an intravenous injection and is carried to cells by the circulatory system. Once it diffuses across the cell membrane, the drug loses its two chloride (Cl) atoms. This loss makes available two *adjacent* sites for binding to cellular macromolecules. Most available evidence points to platinum binding to DNA as the chemical event responsible for the antitumor activity of cisplatin. When platinum binds to DNA in the nucleus of the tumor cell, DNA synthesis is inhibited, the cells cannot divide, and growth of the tumor is arrested.

Interrupting DNA synthesis is the goal of numerous antitumor and antiviral drugs. This strategy works because tumor cells divide quite rapidly, while most cells in adults are not dividing or dividing at a fairly slow pace. The negative side effects associated with cancer chemotherapy are due in part to the fact that these drugs also interrupt the activity of normal cells that need to divide to carry out their functions.

Experiments designed to map the regions of DNA to which cisplatin binds revealed a propensity for the drug to attach itself to two adjacent components in the same strand of the DNA molecule. DNA is composed of two strands coiled in a double helix. Each strand is made up of a linear arrangement of nucleotides, each of which consists of a phosphate group, a deoxyribose sugar moiety and one of four bases—guanine, adenine, cytosine, or thymidine. These bases pair up with bases on the second strand to give DNA is characteristic "ladder" structure. Cisplatin, minus its two chloride atoms, binds to two adjacent guanosine nucleosides or to adjacent adenosine and guanosine nucleosides. The exact positions of binding were located with precision by magnetic resonance and X-ray diffraction structural techniques.

Figure 1 shows how cisplatin binds to its most common target in DNA. As can be seen, the two adjacent sites formerly occupied by the chloride atoms of cisplatin are now used to link the adjacent guanosine nucleosides of the DNA molecule (4). Space-filling views of this structure are displayed in Figure 2. What has been discovered is that such an intrastrand cross-linked structure can occur without grossly disrupting the the DNA double helix, although the helix axis bends by ~40° at the site of platinum attachment (Figure 3). The cross-linking has disastrous, from the tumor cell's point of view, consequences on DNA synthesis.

Studies of the biologically inactive molecule, *trans*-DDP, revealed that, unlike cisplatin, it is geometrically unsuitable for linking adjacent nucleotide building blocks in DNA. Instead, *trans*-DDP prefers to form intrastrand

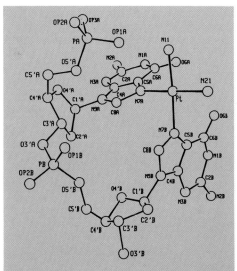

Figure 1. Molecular structure of cisplatin bound to d(pGpG).

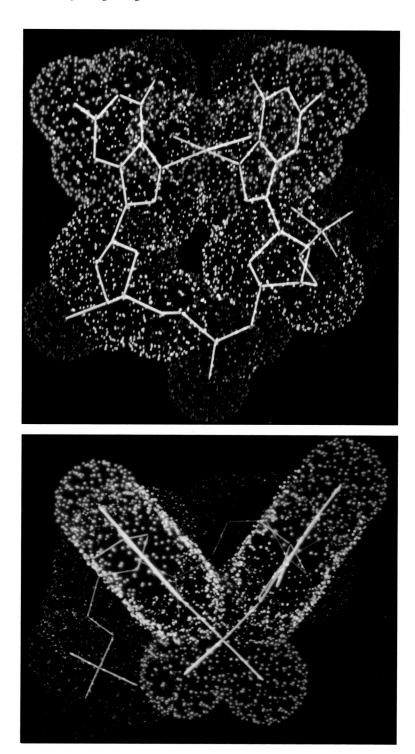

Figure 2. Two views of the van der Waals spheres of the structure in Figure 1. The view on the bottom reveals the hydrogen bond formed between the terminal 5'-phosphate and coordinated ammine groups as interpenetrating van der Waals spheres. (Reproduced with permission from reference 4. Copyright 1985 American Association for the Advancement of Science.)

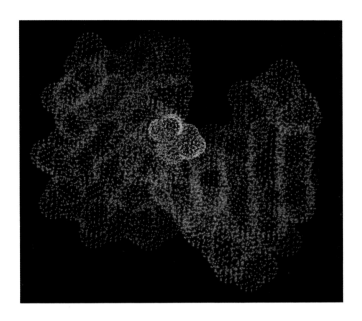

Figure 3. Computer graphics simulation of the binding of cisplatin to its putative target, d(GpG), in a bent double-helical DNA fragment as determined by molecular mechanics calculations.

cross-links with one or more intervening nucleotides (*5, see* structure). In this proposed structure, the intervening nucleotide is unable to pair up with its partner on the opposite strand. In other words, the "rung" of the DNA ladder formed by the intervening nucleotide and its partner is "broken" by the binding of *trans*-DDP. As a consequence, the DNA double helix is likely to be disrupted to a much greater extent following binding by *trans*-DDP compared to binding by cisplatin. This expectation has been verified by experiments in our laboratory (*6*).

Binding of *trans*-DDP and DNA

How might these grossly different structures formed in the chemical reactions of cisplatin and *trans*-DDP account for their different anticancer activities? One might expect that if disrupting a tumor cell's DNA a little bit is a good thing, then disrupting it a lot would be even better. But that is not necessarily the case because tumor cells, like normal cells, have mechanisms for repairing damaged DNA.

Possibly, much higher levels of *trans*-DDP are required to produce an equivalent toxic effect on tumor cells because its removal might be mediated by a cellular repair system that recognizes the perturbations in the DNA structure caused by binding of *trans*-DDP (7). These levels are generally too high to be tolerated for chemotherapy. The toxicity to normal cells is simply too high. Thus, selectivity at the level of molecular recognition by the target molecule might be translated into selectivity at the level of repairing DNA damage. In other words, how the tumor DNA reacts with a given metal compound based on the geometry of that metal compound may translate into selectivity based on a differential ability of the cell to repair damage to the DNA caused by the reaction.

Such a hypothesis, which is currently being tested experimentally, provides the basis for the rational design of other metal compounds that might function with even more selectivity than cisplatin and serve as even more powerful anticancer drugs. A related hypothesis suggests that selective repair of DNA damage by normal tissues and/or by tumor cells could account for the specificity of DNA-damaging agents such as cisplatin. That is, the specificity of such agents for a tumor cell versus a normal cell may be due to the fact that the DNA repair mechanisms function differently in the two cell types. As more knowledge of such phenomena accumulates, and as we learn more about the molecular recognition process of DNA-binding and DNA-damaging molecules, chemotherapeutic agents targeted at specific sequences on the genome will be able to be synthesized by chemists in a powerful attack on genetically controlled diseases.

Biotechnology and AIDS

Let's turn to the newest virally transmitted disease, AIDS (acquired immune deficiency syndrome). Within a few years, it seems to be a certainty that each of us will know personally someone infected with HIV (human immunodeficiency virus). While immunology and epidemiology are the logical best choices for containing and eventually eliminating this modern plague, AIDS will continue to be manifest in those already infected. The numbers of afflicted individuals will be staggering. Estimates from the U.S. Centers for Disease Control in Atlanta suggest that between one and two million Americans already are infected with HIV. Estimates of how many of those people will develop AIDS range from 10% to all of them.

Biotechnology has been essential in developing an understanding of HIV and how it attacks the body's immune system. Much remains to be learned, of course, but the knowledge that has been generated so far is remarkable. As we shall see, chemistry is playing a central role in developing therapeutic agents for AIDS.

HIV is what is known as a *retrovirus*. Most viruses and all other living creatures use DNA as their genetic material. Indeed, it is the universality of DNA as the basic genetic material that makes recombinant DNA techniques and biotechnology possible. By contrast, retroviruses like HIV contain RNA as their genetic material. RNA is quite similar to DNA. The principal differences are that RNA uses the sugar ribose in place of deoxyribose, it uses a base called uracil in place of thymidine, and it is single stranded rather than double stranded. The most salient common feature is that both molecules are capable of carrying the information necessary to produce proteins. That, in fact, is RNA's function in most organisms—to carry the code from the DNA in the nucleus to the cellular machinery that assembles proteins.

Figure 4 shows schematically the life cycle of HIV involving its primary target cell, the helper/inducer T lymphocyte. Although it has now become almost certain that HIV also infects other cells in the body, much of the devastation of AIDS can be explained by the inexorable depletion of these crucial immune system cells as a result of HIV infection. A critical element in this life cycle involves an enzyme called reverse transcriptase. All retroviruses produce a version of this enzyme. In fact, they are all quite similar proteins. Reverse transcriptase catalyzes

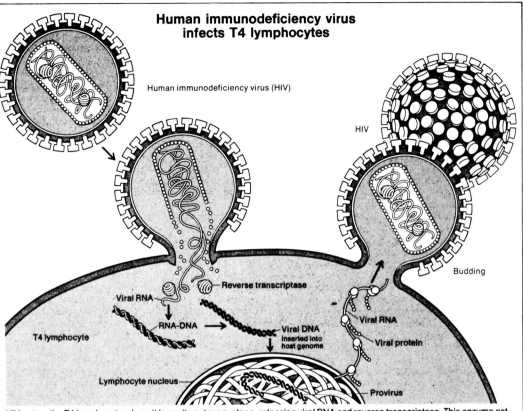

Human immunodeficiency virus infects T4 lymphocytes

Human immunodeficiency virus (HIV)

HIV

Budding

Reverse transcriptase

Viral RNA

RNA-DNA

Viral DNA inserted into host genome

Viral RNA

Viral protein

T4 lymphocyte

Lymphocyte nucleus

Provirus

HIV enters the T4 lymphocyte where it loses its outer envelope, releasing viral RNA and reverse transcriptase. This enzyme catalyzes synthesis of a complementary DNA strand from the viral RNA template. The DNA helix then inserts into the host genome, where it is known as the provirus. When this is transcribed by the infected lymphocyte, possibly after the cell has been activated by exposure to antigen, new viral RNA and proteins are produced to form new viruses that bud from the cell membrane

Figure 4. Life cycle of HIV.

the synthesis of viral DNA from viral RNA after the virus has infected a lymphocyte. (The usual flow of genetic information is from DNA to RNA in a process called transcription; hence this enzyme catalyzes "reverse" transcription.)

Because the reaction catalyzed by reverse transcriptase is an essential process in the HIV life cycle, the enzyme is a target for drugs that might interfere with it and thus disrupt the process of HIV infection. This target is particularly attractive for such a strategy because human cells do not carry out the reverse transcriptase biochemistry. Thus an agent that disrupts reverse transcriptase activity at least potentially is less likely to disrupt normal cell function.

AZT Interrupts

The strategy forms the basis of the mechanism of 3'-azido-3'-deoxythymidine (AZT, *8*), the only anti-AIDS drug so far approved by FDA. Notice how similar the structures of AZT and the thymidine nucleoside are. The azido group, however, which consists of three nitrogen atoms in what is called the 3' position of the sugar molecule, confers an important property on this molecule. Viral reverse transcriptase mistakes AZT for a thymidine nucleoside—in fact, some evidence suggests that the viral enzyme actually prefers AZT over the thymidine nucleoside—and incorporates the AZT molecule into a growing viral DNA chain. The hydroxyl (OH) group at the 3' position of the thymidine nucleoside, however, is essential for attachment of the next nucleoside in the DNA strand. The azido group (N_3) cannot participate in this linkage. With the OH group missing, chain growth is terminated and the viral life cycle is interrupted.

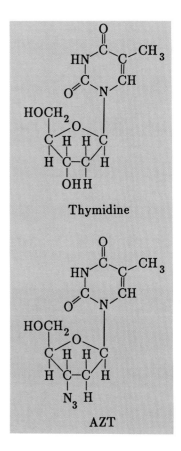

Thymidine

AZT

It turns out that as the first strand of viral DNA is being synthesized by the action of reverse transcriptase, the viral RNA is being degraded and replaced with a second strand of DNA. Thus, when AZT terminates chain growth, the virus is out of luck—its own highly efficient machinery precludes it from starting over again from scratch.

AZT provides another example of how difficult it is to anticipate the practical benefits of basic research. This compound was synthesized more than 20 years ago. Its antiretroviral activity was demonstrated more than 12 years ago. But its use in prolonging the lives of AIDS patients is its first medical application.

Targets for Chemists

The numerous stages in the process by which HIV infects and kills T lymphocytes—binding to cells, entry into cells, transcription of viral RNA to DNA, integration of viral DNA into the cell's DNA, expression of viral genes, assembly of virus particles, and viral budding—all offer targets for chemists to intervene and stymie the virus. Already we know of two proteins, called "tat" and "art/trs", produced by HIV that are essential for viral

*AZT crystals photographed under
polarized light.*

replication. Genetic engineers have introduced these viral genes into bacteria to obtain large amounts of the proteins for characterization. Once we know their structures and how they function, perhaps agents to block their activity can be developed.

There seems to be no better current example of how chemistry can serve society: through the design and synthesis of drugs based on knowledge of the molecular structures of the HIV viral components, processing enzymes, and progeny. As the previous examples have shown, we are moving rapidly toward the day where serendipity will give way to rationality based on knowledge provided by biochemistry and molecular biology. We chemists stand ready to meet this challenge through the molecular basis of drug design.

References

1. Perutz, M. F. "Ging's ohne Forschung besser? Der Einfluss der Naturwissenschaften auf die Gesellschaft"; Wissen-

schaftliche Verlagsgesellschaft mbH: Stuttgart, West Germany, 1982.

2. Rosenberg, B. In *Nucleic Acid–Metal Ion Interactions*; Sprio, T. G., Ed.; Wiley: New York, 1980, p 1.

3. Lippard, S. J. *Pure Appl. Chem.* **1987,** *59,* 731.

4. Sherman, S. E.; Gibson, D.; Wang, A. H.-J.; Lippard, S. J. *Science* 1985, *230,* 412.

5. Lepre, C. A.; Strothkamp, K. G.; Lippard. S. J. *Biochemistry,* in press.

6. Sundquist, W. I.; Lippard. S. J.; Stollar, B. D. *Biochemistry,* **1986,** *25,* 1520.

7. Ciccarelli, R. B.; Solomon, M. J.; Varshavsky, A.; Lippard. S. J. *Biochemistry,* **1985,** *24,* 7533.

8. Mitsuya, H.; Broder, S. *Nature,* **1987,** *325,* 773.

Chapter 5

Taming the Chemistry of Proteins

Emil Thomas Kaiser

A revolution has occurred in our understanding of proteins, the biological macromolecules composed of amino acids that are crucial to all aspects of life from breathing to digestion to vision. This revolution is an essential, although sometimes overshadowed, element of the advent of biotechnology. Genetic engineering so often is regarded as synonymous with recombinant DNA, and of course, this remarkable technology is the basis for much of biotechnology. However, one of the principal reasons that scientists manipulate DNA is to create new proteins. DNA, after all, exists in large measure to provide a template for the production of proteins. Without a detailed understanding of protein structure and function, scientists would often not have a rationale for their manipulations of DNA. And that understanding is the product, primarily, of almost 40 years of work by chemists and biochemists.

The Structural Chemistry of Proteins

Proteins are, in their most simple guise, a linear arrangement of amino acids linked together with what are called amide bonds. Proteins can contain hundreds of amino acids, but there are only 20 different amino acids for organisms to choose among to build the enormous array of proteins that exist in nature. The linear sequence of amino acids in a protein is referred to by scientists as the *primary* structure of the protein. If we look back only three decades, which is when I was beginning to become interested in proteins, we find that the primary structure of only one protein—insulin—had been determined. Figure 1 shows this structure.

1473-5/88/0043$06.00/0

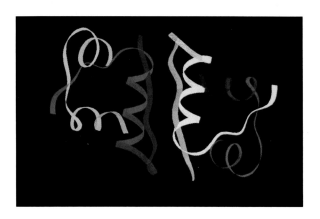

```
                             S———————————————————S
                             |                    |
H–GLY–ILE–VAL–GLU–GLU–CYS–CYS–THR–SER–ILE–CYS–SER–LEU–TYR–GLN–LEU–GLU–ASN–TYR–CYS–ASN–OH–
                             |                                                  /
                             S                                                 S
                             |                                                 |
                             S                                                 S
                             |                                                 /
H–PHE–VAL–ASN–GLN–HIS–LEU–CYS–GLY–SER–HIS–LEU–VAL–GLU–ALA–LEU–TYR–LEU–VAL–CYS–GLY–GLU–

ARG–GLY–PHE–PHE–TYR–THR–PRO–LYS–ALA–OH
```

Figure 1. Top: The primary structure of insulin. Bottom: A "ribbon" structure of the insulin dimer.

Sanger Determined the Primary Structure of Insulin

This seminal achievement by Frederick Sanger occurred through the application of new organic chemical methodology. The work required six years to complete, and it was recognized in 1958 with the award of Nobel Prize in Chemistry. Until Sanger's efforts, the idea of trying to unravel the primary sequence of a protein was one that was too daunting to contemplate. He chose bovine insulin because it is a relatively short protein, containing only 51 amino acids, and because large quantities were available to him for analysis. Sanger found that bovine insulin consists of two chains, one containing 30 amino acids and the other 21 amino acids. This finding made the problem of sequencing the protein somewhat easier, because Sanger was able to sequence the chains individually. To oversimplify, Sanger used a chemical that reacts with the amino acid at the end of the polypeptide chain to produce a strongly colored derivative of that amino acid. He then broke the polypeptide into individual amino acids and identified the strongly colored derivative as the first component in the chain. By a lengthy series of reactions, Sanger was able to reconstruct the sequences of the two chains that make up insulin.

Sanger's work on insulin was followed by several major research successes. Among these were the determination of the primary structure of other types of proteins such as ribonuclease, an important enzyme that digests RNA. Proteins, however, possess structure beyond merely the linear sequence of amino acids that make them up. In aqueous solution, proteins possess a complex architecture that can be categorized in a hierarchy of what are called secondary, tertiary, and even quaternary structures. In the years following Sanger's work, some of these levels of structure began to be unraveled for different proteins, largely through the application of X-ray crystallographic techniques.

Pauling Deduced the First Secondary Structure

In 1951, for example, Linus Pauling worked out the first *secondary* structure for proteins. Pauling focused on determining exact values of bond lengths and bond angles for the primary bond in proteins, the peptide bond that links one amino acid to the next amino acid. To make this determination, he used X-ray crystallography to measure these values in small, model peptides. These experiments led him to propose the idea of an α-helix as a basic unit of secondary structure in proteins—amino acids coiled in a right-handed spiral held in place by well-defined hydrogen bonds. Very few proteins, however, are entirely in the α-helix form because other factors can disrupt this element of secondary structure. Other secondary structures also exist. Pauling received the 1954 Nobel Prize in Chemistry in part for his seminal work on protein structure.

Tertiary and Quaternary Structure

Proteins, researchers discovered, consist of domains of secondary structure that fold under the influence of other forces into precise three-dimensional configurations—the tertiary structure—that influence greatly the function of the protein, especially if the protein acts as an enzyme. In some cases (the critical oxygen-transporting protein hemoglobin is a perfect example) more than one of these globular protein units come together—the quaternary structure—to form the functional protein. In the case of hemoglobin, four chains, two designated α and two β, allow the protein to bind to oxygen under certain physiological conditions and release it under other physiological conditions.

In parallel with these successful efforts to understand the structural aspects of proteins, the molecular architecture of DNA was determined in the brilliant description of the double helix by Watson and Crick, an effort for which they won the 1962 Nobel Prize in Medicine. Subsequently the genetic code was broken, allowing scientists to perceive the relationship between the linear sequences of the basic building blocks in DNA—the A, G, C, and T nucleotide components that have been discussed in previous chapters—and the amino acid components of proteins.

With these formidable achievements in structural chemistry, the stage was set for the leap into the new era of biotechnology.

Sequencing DNA and Proteins

With these formidable achievements in structural chemistry, the stage was set for the leap into the new era of biotechnology. Before that leap could be made, however, further advances were required. Although the linear sequences of the components of proteins and nucleic acids had been determined in a number of cases and the relationship between these sequences recognized, the methodology that had been used was painstakingly slow. This drastically limited the number of systems that were amenable to study. Before one is able to engineer nucleic acids or proteins, one must have the fundamental information about those molecules—the nucleotide or amino acid sequence—in hand.

Early successes in determining the sequence of nucleotides in stretches of DNA, for instance, were based on an indirect method of sequencing that involved transcription of the DNA into mRNA and subsequent sequencing of the mRNA. Not only was this method slow, but it was also prone to errors. Direct methods of sequencing DNA were developed by Walter Gilbert and A. Maxam, who used chemical reagents to cut DNA molecules, and by Frederick Sanger, who used enzymatic methodology. The development of these methods was recognized with the award of the 1980 Nobel Prize in Chemistry to Gilbert and Sanger, who shared it with Paul Berg, a central figure in the development of recombinant DNA techniques.

The Maxam–Gilbert Method

The Maxam–Gilbert method of sequencing DNA provides a vivid illustration of the chemical ingenuity that has been applied to these biological problems. A feature of DNA, based on how the sugar groups of nucleotides are linked together through phosphodiester bonds, is that the two strands of DNA run in opposite directions—they are what is referred to as "antiparallel". Because of this, the ends of DNA strands are distinguishable from each other, and are referred to as the 5'- and 3'-ends. The Maxam–Gilbert sequencing technique uses restriction enzymes to digest the double-stranded DNA to be sequenced into fragments about 100 base pairs long. The 5'-ends of these fragments are labeled with radioactive phosphorus, and the double-stranded DNA fragments are separated into single strands. By using the technique of chromatography, these strands are separated into fractions, and each fraction represents one particular DNA fragment to be sequenced.

Each fraction is then divided into four portions, and the portions are subjected to one of four chemical reactions. One reaction cuts the DNA strand at adenine, another at guanine, another at cytosine, and another at thymine. Because Maxam and Gilbert allowed only an incomplete digestion of the fragments—that is, only one or two cuts per strand—this method provides a mixture of fragments of different lengths ranging from, in this example, 1 to 100. Not all fragment lengths are represented in each portion; only those corresponding to the base for which the chemical reaction was specific. Therefore, the same base is at the 3'-end of every fragment from a given portion.

The fragments are then separated by a technique called "gel electrophoresis" in which fragments of different lengths migrate at different speeds in a gel. The result is a series of bands, which can be identified because of the radioactive phosphorus, corresponding to DNA fragments of different lengths. The length of each corresponds to a position occupied by a given base in the DNA. The four different chemical reactions, and four different electrophoresis patterns, provide the complete sequence of that stretch of DNA. As complicated as it sounds, this technique was an enormous improvement over the previous sequencing technique, and it could be, and subsequently was, automated.

Edman's Technique

Similarly, techniques were needed to sequence proteins in a fashion more rapid than that developed by Sanger for his work on insulin. This improvement was accomplished by Per Edman in the late 1950s. He developed a technique in which a chemical called phenyl isothiocyanate is added to the amino acid at one end of a protein. This molecule destabilizes the peptide bond that holds that amino acid to the amino acid next to it in the chain, and the bond breaks. The free amino acid's identity is determined, and the procedure is repeated again and again until the entire protein has been digested. This procedure also has been automated. Today, the protein-sequencing machines based on Edman chemistry need less than a microgram of an average protein on which to carry out a complete series of sequencing reactions. It is now possible for a scientist to put a protein into the reaction vessel of one of these machines, come back in a few hours, and find a printout of the linear sequence of amino acids in the protein.

Synthesizing Proteins and DNA

To make this story complete, an additional development that was required to fuel the growth of biotechnology was the ability to synthesize proteins. Two major methodologies have been developed to accomplish this step. One, which is entirely dependent on chemical synthesis, was devised by my colleague at the Rockefeller University, R. Bruce Merrifield. In this technique, proteins are assembled by repetitive coupling of the constituent amino acids attached to a polymeric support. This procedure also has been automated. It greatly increases the rapidity of protein synthesis and permits preparation of considerable quantities of proteins. An important feature of the technique is that it allows the building of proteins that contain amino acids other than the 20 used by living systems. Considering the other scientists mentioned in this narrative, it should come as no surprise that Merrifield received the 1984 Nobel Prize in Chemistry for his research.

The other technique for building proteins is the one that much of this book is about: genetic engineering. Proteins can be produced by genetic engineering in several ways. In many of them, chemical synthesis methodology conceptually related to the solid-phase synthesis of proteins—the Merrifield technique—is used to prepare synthetic DNA that is critical in the early stages of the work. So we see that four chemical technologies are central to biotechnology:

- protein sequencing
- nucleic acid sequencing
- protein synthesis
- nucleic acid synthesis

Relationship Between Structure and Function

In my laboratory, we are making efforts to understand a number of aspects of the relationship between structure and function in proteins. For example, we studied the role of secondary structure in a bacterial enzyme called alkaline phosphatase. Part of this protein called the signal sequence has a 21-amino acid primary structure that has a hydrophobic core region thought to take on an α-helix secondary structure. This signal sequence is needed for the enzyme to be transported across the bacterial cell membrane. We devised a different sequence in the core region based simply on having a repeating pattern of leucine residues that bore little relationship—or, in the language of molecular biology, sequence homology—to the natural sequence, except that one would expect our sequence to show an even greater tendency to take on an α-helix secondary structure. Using the techniques of genetic engineering, we altered the bacterial genome so that it would produce an alkaline phosphatase containing our signal sequence (Figure 2). What we found is that the enzyme is processed normally by the bacteria and, indeed, more rapidly than observed for the natural signal

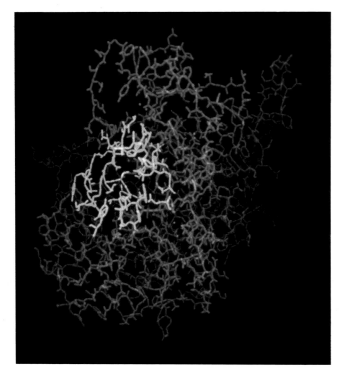

Figure 2. A computer graphics representation of the monomeric structure of thiol alkaline phosphatase, a mutant enzyme.

sequence. This finding suggests that the secondary structure itself is what is essential for the function of this particular signal sequence.

In another approach to the development of a better understanding of structure–function relationships, we have pursued the "chemical mutation" of enzyme active sites to give new catalytic species. We have alkylated the active site residue cys-149 in glyceraldehyde-3-phosphate dehydrogenase (GAPDH) with a flavin analog, a compound with a ring structure related to that of riboflavin. The resultant flavo-GAPDH, a semisynthetic enzyme, functions as an efficient oxidoreductase, which uses the flavin group as its reactive function in the oxidation of the important coenzyme NADH. A picture of the active site of flavo-GAPDH with a bound NADH substrate molecule is shown in Figure 3. One of the exciting features of the chemical mutation strategy is that drastic alterations of the catalytic activity of enzymes are possible. We have shown that the hydrolytic enzyme papain can be turned into an effective oxidoreductase by appropriate modification with a flavin, for example.

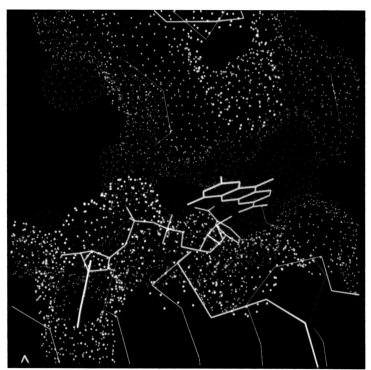

Figure 3. Closeup of the active site of the semisynthetic enzyme flavoglyceraldehyde-3-phosphate dehydrogenase with a bound substrate (NADH) molecule.

Fewer Limits, Dreams Realized

Through these tremendous contributions from chemistry, the tools for the construction of proteins have become available. Starting with targets like genetically engineered insulin, we have moved to the point where the availability of pharmaceutically important proteins is no longer limited by the requirement to isolate them from the natural sources. Proteins now can be made as they are needed. Despite these dizzying achievements, the greatest period for the application of chemistry to biological problems lies in the future. Chemistry brings a unique perspective to the question of the relationship between structure and function. It has now become possible, by using what is known about protein architecture, to redesign protein molecules with the goal of targeting them for specific biological tasks. We can improve their stability and enhance their biological activity. We have entered into an era where we shall be able to establish control of protein function through systematic alteration of protein structure. It is already clear that we are rapidly moving toward the time when we may use the amino acid building blocks of proteins to design entirely new protein molecules for specific biological and chemical functions. Researchers already are taking the initial steps to realize that dream.

The impact of efforts to design proteins is already extremely broad. The targets include proteins of agricultural, medical, and industrial importance. The agricultural proteins range from growth-regulating factors to insecticides. The medically important proteins include a host of hormones, antibodies, vaccines, and therapeutic enzymes that control the most fundamental biological processes such as blood clotting. The use of enzymes in industrial processes—which holds the potential of cutting dramatically the energy needed for such processes—is also being enhanced through redesign efforts that allow alterations to the enzymes thereby producing, for example, improved heat and pH stability.

When I reflect on where the application of chemistry to proteins has taken us, I cannot help but express my feeling that the period ahead of us is a very special one. For a scientist at the present time, the excitement of being able to tame the chemistry of proteins could have its parallel in the exhilaration felt by an artist during the Renaissance.

Chapter 6

Protein Engineering

Gregory A. Petsko

In April 1987, a meeting with the simple title "Protein Engineering" was held in Oxford, England. The organizers expected 50 to 100 people to attend, but more than 500 came. Another meeting with the same title was, coincidentally, held at exactly the same time in Colorado. Almost 500 people came to that one, too. The subject of these two meetings—protein engineering—did not exist five years ago.

In 1986, a new scientific journal called *Protein Engineering* was started. The publishers received more subscriptions than for any start-up journal in their history before a single issue had yet been published.

The Japanese government is spending $50 million over the next three years to build a Protein Engineering Research Institute near Osaka University and is staffing it with the finest scientists culled from institutions across the country. Japanese industry will probably contribute an equal amount of money to this venture.

What's all the fuss about?

The Second Chemical Revolution

As has been pointed out in previous chapters, we are at the beginning of a revolution. It is actually the second chemical revolution. The first chemical revolution occurred more than 100 years ago when chemists learned how to make simple organic molecules by mixing together even simpler molecules in test tubes. That revolution freed humans to an extent from the need to extract chemicals from plants or distill them from vats of yeast. It allowed humans to create chemicals other than those that Nature provided. That revolution began with urea and ended with nylon. In between came chemical discoveries that changed the world: gasoline, synthetic

1473-5/88/0053$06.00/0

penicillin, plastics of all types, synthetic rubber; the list is almost endless.

The second chemical revolution began just a few years ago when chemists learned how to manipulate the chemical building blocks of living cells to produce totally new genes and proteins. The industry that this revolution spawned is called *biotechnology*, but it is really applied biological chemistry. It depends on a set of rules about the way that the three-dimensional structure of a biological molecule influences its chemical properties. One part of this science is called *protein engineering*: redesigning proteins to carry out new functions. This seems like a tall order. Proteins are very small and very complicated, and the parts of proteins that scientists want to tinker with— amino acids—are smaller still. However, we can do it. We could not do it five years ago, and we could not even imagine how to do it 10 years ago; but we can do it now. In fact, as earlier chapters have pointed out indirectly, we can do it quite easily. That is what all the fuss is about.

Protein engineering has as its goal the design and creation of improved proteins. Ralph Waldo Emerson anticipated the reason for the industrial interest in protein engineering when he wrote: "Let any man build a better mousetrap, and the world will beat a path to his door." Overlooking the sexism of this remark—Emerson was, after all, living in a sexist age—one can assert that his comment still applies. If a person builds a better mouse-trap, the world will still beat a path to his or her door. In protein engineering, the mousetraps are proteins.

As has been pointed out in previous chapters, proteins are linear polymers of amino acids strung together like beads on an open necklace. Each necklace may contain hundreds of beads, but they all come from a basic library of 20 different types. Depending on how many beads there are in the necklace, the number of each type used, and the order in which they are arranged along the string, an enormous number of different necklaces can be made. Such an enormous number, in fact, that all the differences among living organisms result from different collections of these necklaces called proteins. Changing one bead in the necklace to a different type produces a necklace that is different, even if only subtly.

Protein engineering is carried out by changing individual amino acids in the protein to one of the other 19 possibilities. Of course, Nature may already have tried

The second chemical revolution began just a few years ago when chemists learned how to manipulate the chemical building blocks of living cells to produce totally new genes and proteins.

Figure 1. Space-filling computer graphics representation of the classic Watson–Crick double helix for DNA. (Photo courtesy of Richard Feldmann, NIH.)

out this possibility earlier in evolution, and discarded it because of pressure from natural selection. In other words, a different choice, maybe the one we find in that protein today, conferred a survival advantage on whatever species had it. Our requirements, however, as industrial chemists or physicians may be quite different from those of the organism that evolved a protein of interest. An industrial chemist working with an enzyme, for example, might be willing to lose a certain amount of catalytic activity in exchange for significantly increased stability in acidic solutions.

Figure 1 shows the by-now familiar structure of DNA, the ladder of life. As we have seen, genes are stretches of DNA that code for protein molecules. The genetic code is in the rungs of the ladder: three consecutive rungs of code for a particular amino acid. Protein engineers use recombinant DNA technology to alter a particular triplet of rungs so that a different amino acid is coded. The process is really microsurgery, performed not with scalpels or lasers, but with chemistry. It is very much like editing text on a word processor. Genetic engineers cut pieces out of the DNA and move them around, or replace one three-letter word—or *codon*—with another carefully chosen word. The result is an altered or mutated gene. When this mutated gene is introduced into bacteria or yeast or a plant or a mouse, the organism will produce the altered protein encoded by the mutated gene. If that protein is meant to work inside the organism, then the effects of the new structure can be studied. This step has been done in the case of mice that received a gene for growth hormone from rats; this genetic surgery produced mice that grew to twice the normal size. Otherwise, the protein can be purified from the bacteria or yeast and used to treat people or produce chemicals.

Designing Protein Properties

What makes protein engineering such a "hot" field is that scientists have developed a good understanding about how to design the specific changes they need to obtain whatever property they want in a protein. This advance in understanding occurred simultaneously with the development of the methods to make the changes. If that had not happened, we would have had to fumble about by trial and error to obtain the new proteins. That is the way evolution works, but Nature has a long time to experiment, and does not have to write grants or answer to stockholders. Because the necklace of a protein molecule is coiled up in space into a definite three-dimensional structure, and because that structure often actually can be determined by X-ray crystallography, scientists can see how the various amino acids in the protein interact with one another to carry out the function of the protein. Also, scientists can predict, with surprising accuracy, what will happen when one amino acid is changed into another amino acid. The result is that we can proceed, not by trial and error, but by design, and produce a better "mousetrap".

A few specific examples of the things that are being done today in laboratories can provide a sense of why there is such excitement about this kind of science in the medical and industrial communities.

Treatment of Emphysema

The human lung requires a very special kind of protein in order to be able to expand and contract as a person breathes. This protein is called elastin, and it is named for its elastic property. When a person's lungs are damaged, by pollution, cigarette smoke, or bacterial infection, the damaged elastin must be cut away so that fresh elastin can be synthesized to take its place. Another protein, an enzyme called elastase, carries out this function. Elastase is similar to the digestive enzymes in the human stomach that chew up the proteins in ingested food. (Nature tends to be conservative; the same biochemical motif appears in enzymes that catalyze similar chemical reactions in widely varied tissues.) Elastase's role is to chew up elastin.

Obviously, it is dangerous to have a digestive protein—a sort of biological Pac Man—running around in the lungs uncontrolled, so Nature has developed yet a third protein, called antitrypsin, that inhibits elastase. Antitrypsin plugs the active site of elastase like a cork in a bottle, as shown in Figure 2. Antitrypsin is removed when the body needs to turn on the elastase to repair damaged lung tissue. In one disease, cigarette smoke (or in some cases a genetic defect) chemically alters a single amino acid in the antitrypsin so that it is physically larger than in the normal protein. The altered protein cannot plug up elastase any longer. When this happens, the

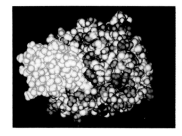

Figure 2. Space-filling model of the complex of a protease, trypsin, with a naturally occurring trypsin-inhibitor protein. The inhibitor is shown all in white; the enzyme is multicolored. This three-dimensional structure was determined by X-ray crystallography by Robert Huber and his associates at the Max Planck Institute in West Germany. Although it is not a model of the elastase–α$_1$-antitrypsin complex, which has not yet been solved by X-ray crystallography, it does illustrate the general principle that protease-inhibitor proteins plug the active site of their target proteases like a key in a lock.

elastase is uncontrolled and the lungs slowly digest themselves. This disease is called emphysema, and it kills thousands of Americans every year.

Protein engineering provides an approach for the treatment of emphysema. Researchers at Transgene, a genetic engineering company in Strasbourg, France, have changed the amino acid that is attacked chemically by cigarette smoke or pollution to one that is inert to this damage. The result is a resistant version of antitrypsin, and clinical efforts are now underway to deliver this modified protein to the lungs of patients with emphysema.

Manufacturing Amino Acids

Amino acids are not just the building blocks of proteins, but are also valuable industrial chemicals. Aspartame, which goes by the trade name Nutrasweet, is made up of two amino acids—phenylalanine and aspartic acid—and there are many other, less familiar examples. Amino acids have been referred to throughout this book as simple molecules, and indeed, they are, relative to the complexity of the proteins they make up. However, amino acids are complex enough that manufacturing a large quantity of a given amino acid in the correct configuration remains a fairly expensive undertaking. Thus, the efficient industrial manufacture of amino acids from cheap starting materials is a major research goal worldwide.

Enzymes called *transaminases* can do precisely this chemistry, and do it very efficiently. The problem with using them commercially is that there are not transaminases that produce all of the desired amino acids. Like most enzymes, the transaminases are very specific—for the purposes of the industrial chemist, too specific—and accept only certain starting materials to produce only certain amino acids. This limitation can be overcome by protein engineering.

Jack F. Kirsch and Wayne Finlayson at the University of California, Berkeley, and Steve Almo, Doug Smith, and Dagmar Ringe at the Massachusetts Institute of Technology examined the structure of the transaminase that binds aspartic acid. They found that the substrate aspartic acid molecule is bound at the active site of the protein by an arginine residue as shown in Figure 3. They reasoned that this situation could be reversed. That is, if the arginine amino acid residue in aspartate transaminase

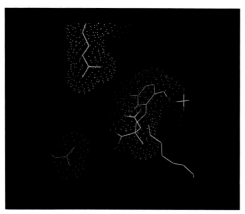

Figure 3. Crystal structure of the active site of Escherichia coli aspartate aminotransferase, deterained by Doug Smith and Dagmar Ringe of MIT. A model of the substrate aspartic acid has been built attached to the pyridoxal phosphate cofactor. In this model, constructed by Steve Almo of MIT, the side-chain carboxylate of the substrate (the two carboxylate oxygens in red at the center of the figure) is making a salt bridge to the guanidino group of arginine 292 at the bottom of the figure. This interaction gives the enzyme its specificity for aspartic acid.

were replaced by an aspartic acid residue, then the new enzyme might bind arginine. This experiment was a success. By a single amino acid change, which is relatively simple to accomplish, they were able to turn aspartate transaminase into arginine transaminase.

It should be possible to apply this methodology to many other systems where the enzyme that exists is not specific for the molecule of interest. In fact, it already has been successful in a number of cases.

Engineering Antibodies

Antibodies are proteins that bind foreign substances (Figure 4). Their function is normally only binding. They do not carry out chemical reactions on the things they latch on to. As pointed out in Chapter 3, however, independent research in the laboratories of Peter Schultz and Richard Lerner successfully used the techniques of protein engineering to turn antibodies into enzymes. Because it is quite easy to raise antibodies against almost any complex chemical, this approach opens up exciting possibilities for the creation of totally new protein catalysts.

Antibodies can also be engineered for improved medical applications. Many biotechnology companies like Centocor and Xoma are focusing primarily on mono-clonal antibody technology rather than recombinant DNA technology. Many other genetic engineering com-panies have research programs in both areas. One exciting approach is to attach toxic protein molecules to anti-bodies and use the selectivity of the antibody molecule to carry the toxin to the desired target.

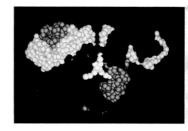

Figure 4. Space-filling model of the three-dimensional structure of im-munoglobulin, determined crystallo-graphically by David Davies and associates at NIH. The molecule has four polypeptide chains: the two heavy chains are shown in blue and red; the two light chains are depicted in yellow. The stem of the T-shaped molecule is the Fc fragment that interacts with cell-surface receptors. The white groups are carbohydrates attached to the protein. The ends of the cross-piece of the T shape are the antigen-binding regions. Many dif-ferent kinds of molecules can be bound by antibodies, but each anti-body generally binds only a few closely related antigens. (Photo by Richard Feldmann, NIH.)

Many of the chemicals used in, for example, cancer chemotherapy take advantage of the fact that tumor cells often are rapidly dividing while most of the cells in adult humans are not. Thus the selectivity of these agents depends on a differential in how fast different cell types are dividing. The compounds, however, remain highly toxic to all dividing cells, and this accounts for many of the side effects associated with chemotherapy. A more promising strategy would be to deliver the toxin only to the cells to be killed (*target cells*).

The protein ricin is a plant toxin obtained from castor beans, and it is one of the deadliest poisons known. A single molecule of ricin may be sufficient to kill a cell. A tiny amount of it on the tip of an umbrella was used to kill a Bulgarian defector a few years ago in London in a real-life espionage story. No doctor would dream of using ricin therapeutically because it kills cells indiscriminately. Like the herbicides discussed in Chapter 2, ricin is almost entirely nonselective.

Research has shown that ricin is composed of two polypeptide chains, designated the A chain and the B chain. The ricin B chain is responsible for binding the protein to target cells. This chain recognizes a common sugar residue that is found on the surface of most mammalian cells, and this recognition accounts for the nonselective nature of ricin. The B chain also facilitates entry of the ricin A chain into the cell. The A chain strongly inhibits the process of protein synthesis by interfering with cellular organelles called ribosomes. Thus the toxic activity of ricin is localized on one part of the protein molecule, the A chain.

Protein engineers are taking advantage of this property of ricin. The idea is to replace the nonselective B chain of ricin with a highly selective monoclonal antibody. If these antibodies, for example, are ones that have been raised to be directed against a type of tumor cell, then the ricin A chain will be delivered only to those cancerous cells, and its toxic action will be specific and beneficial. Scientists at Cetus, a biotechnology company in California, have shown that this approach allows selective killing of breast cancer cells in test animals. These researchers plan to test the compound in human clinical trials. Scientists at other companies are pursuing similar research. Such antibody-toxin constructions are called *immunotoxins*, and there is great excitement about them in the medical community. They offer a whole new approach in the treatment of many diseases besides cancer.

Glucose Isomerase

Finally, the protein shown in Figure 5, glucose isomerase, is the most widely used industrial protein in the world. It catalyzes a reaction that converts glucose, which is not very sweet, into fructose, which is. Every time you drink a Coke or a Pepsi that is not a diet soft drink, you are drinking something that has been sweetened with the product of glucose isomerase. The annual market for this enzyme—the enzyme itself, not its product—is over $70 million per year. Engineering a glucose isomerase that is more stable to heat, or to acid, or that worked faster, would have a major impact on a very large industry. This three-dimensional structure, which was solved at Massachusetts Institute of Technology by Greg Farber, Dagmar Ringe, and me, is the first step in that direction.

Figure 5. Computer graphics representation of the backbone folding of the enzyme glucose isomerase, the enzyme that is used industrially to manufacture high fructose corn syrup.

The Threshold of the Revolution

There are dozens of other examples of the power of this technology to rationally design proteins. Enzymes used in laundry detergents have been improved. Protein engineering has produced new adhesives that work in salt water. Plant growth has been improved by eliminating a wasteful reaction. Dead protein fragments have been produced for use as a vaccine against hepatitis, and the same strategy is being applied to AIDS and other diseases. A number of scientists are close to being able to build a functional protein from scratch. The list goes on, and we are only at the threshold of the revolution.

This is the most exciting time in the history of science to be doing chemistry, and chemistry is the most exciting science to be doing. Scientists can now do transplant surgery on the stuff of life itself. We are actually building better mousetraps—starting, of course, with some pretty good ones provided by Nature.

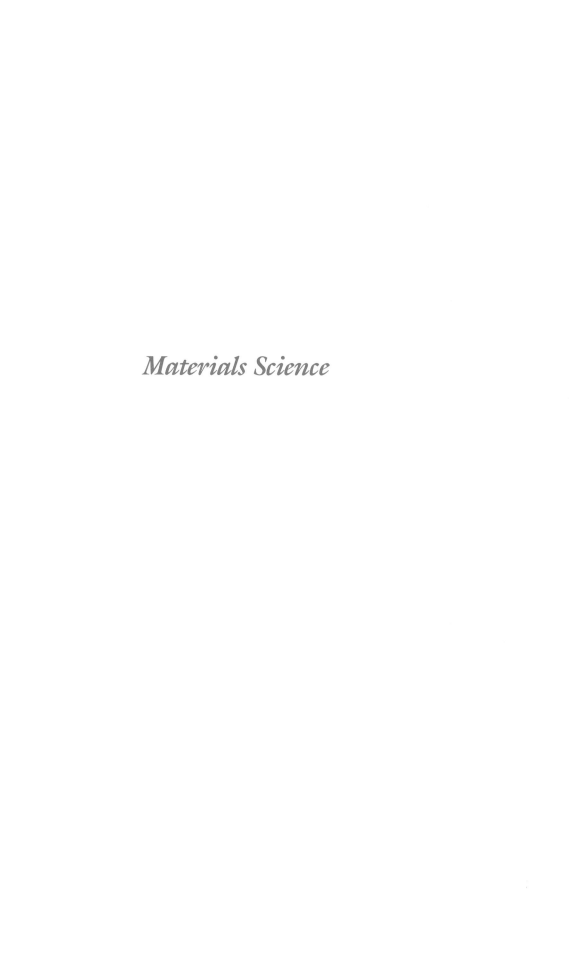

Materials Science

Chapter 7

Chemical Research in Materials Science

William P. Slichter

Materials science is a newcomer in the lexicon of science. This fact is puzzling in the history of human events, and may only reflect the language styles of our pragmatic times. But materials as entities of civilization, especially in modern times, are intrinsically descriptive of our life style. Indeed, a clear measure of the emergence of humans from their precursors was the evolution of crude tools made of stone. (They probably began with organic materials, which of course have not survived).

In comparatively recent times, maybe 10,000 years ago, the disciplines of materials began to appear upon the human scene. These disciplines included fuel refinement, smelting, alloying, and the firing of ceramics. The arts of metallurgy, such as those governing weaponry, were often sophisticated but always empirical. Tanning and dyeing involved advanced chemistry, even though they were only crafts.

Only in the middle of this century has the term "materials science" come into explicit use. Much of the content of the term is associated with technologies that, in present times, invoke a combination of the familiar sciences. New emphasis on materials has arisen from visionary support by the federal government of new research activities, especially under the Advanced Research Projects Agency (ARPA) of the Department of Defense, the National Science Foundation, and the Department of Energy.

So, although materials science as a collection of disciplines in service to technology is ancient, its history as a broadly organized endeavor is rather new. Materials science deals with substances and the processes that make them useful. Traditional disciplines of physical science—

1473-5/88/0063$06.00/0

chemistry, physics, metallurgy, ceramics—apply centrally to materials science, and are joined by newer specializations such as semiconductor physics, nuclear chemistry, and biotechnology. Each of these, and others, brings many of its advances into union with materials science. In consequence, materials science is a multidisciplinary activity, in which chemistry has special prominence.

Chemistry and Materials

Materials science mingles the broad disciplines of synthesis, characterization, and processing. Chemistry is not the sole proprietor of these subjects but it is a major participant in each.

Chemistry, since the early 19th century, has been the main source of information for the synthesis of new materials and for controlling their compositions and properties. Chemistry intrinsically provides the talents to fabricate new substances and to understand what has been made—its composition, structure, purity, and reactivity.

Synthesis

Synthesis has always been a frontier of chemistry. Even the Renaissance included the beginnings of synthesis, but only as a trial-and-error empiricism. The advent of synthetic chemistry beyond the level of alchemy involved isolating elements or simple substances found in nature and combining them in measured proportions. The techniques of physical measurement needed to characterize what had been put together were developed as necessary tools. The philosophy and the methods have advanced tremendously in the two centuries that mark the passage into modern synthetic chemistry, but the formation of new matter remains a central theme of chemistry.

An early, important example of the application of synthetic chemistry to the production of materials was the synthesis of fabric dyes from compounds extracted from coal tar, which occurred in the middle of the 19th century. Prior to the discoveries of those processes, all dyes were substances that occurred naturally. The chemical constitutions were largely unknown, and direct synthesis from basis chemicals was hardly dreamed of. The first dyestuffs were made purely empirically, but the growth of the theory of molecular structure and the theoretical approach to reactions led to the beginnings of synthesis as an exact materials science.

The power of chemical synthesis in materials science and technology has become enormous. Although mining and forestry are still major sources of materials, chemical synthesis gives us access to a boundless array of compounds not found in nature, and also provides economically attractive routes to a host of natural substances using synthetic processes.

A classic example of the power of synthesis is in the domain of rubberlike materials. Natural rubber, the product of the rubber tree *Hevea brasiliensis*, has been used in automobile tires and waterproofed fabrics for many decades. When World War II interrupted the Asian supply of natural rubber to the United States, American chemists were able, with great effort, to invent and develop a family of synthetic rubbers that satisfactorily answered the acute demands for critical materials. Synthetic rubbers proved to be much more than mere substitutes. Synthetic and natural rubber live today in a harmony that is governed by economics and the optimization of properties.

A host of modern synthetic methods has allowed us not only to copy the molecules of nature but also to invent a huge variety of substances that nature had not thought of. The fibers used for textiles are important examples. The classic materials are cellulosic (cotton, flax, and hemp) or proteinaceous (wool, silk, and furs). Synthetic chemistry has brought forward the polyamides (nylons), polyesters, polyacrylonitriles, and polyolefins that have overwhelmed the natural fibers for a great many uses. Moreover, synthesis has provided us with special fibers with properties not afforded by nature, for example, the fireproof aromatic polyamides and the refactory carbon fibers. These substances are based upon petroleum as a raw material and therefore are basically cheap.

These synthetic materials are examples of the strength of synthetic chemistry in competing with nature in materials of high demand. Chemistry is equally important in the synthesis of a myriad of materials *not* found in nature. Examples of these are such exotic

Chemistry intrinsically provides the talents to fabricate new substances and to understand what has been made. . . .

specialties as adhesives that vastly outperform common glues, insulators of extreme electrical strength, photo-chemicals that respond swiftly to specific wavelengths in color photography, chemical preservatives that vastly extend the useful life of other materials, and ceramic composites that will function in place of metals in automobile engines. Such substances come directly from chemical research, and are not mere refinements of lesser materials.

Synthetic chemistry continues to offer high promise for new materials that will excel those that now exist. Every area of materials science has its examples of opportunities that could open up new industries. Simple examples include light-switching materials that lend the speed of light to the operation of computers; new generations of composite fiber-containing materials for structures such as aircraft that must withstand ever more extreme temperatures; and practical superconducting materials for electric power networks.

But the foregoing materials, though not yet realized in terms that engineers can make use of them, neverthe-less already exist in prototype form. The big challenge for chemistry, as the core of materials science, is to find substances possessing properties or performance that invade the unknown.

Chemical Kinetics

Chemical change is the partner of chemical synthesis in the field of materials science. The rates at which change occurs in materials are central to both the formation and the useful service of substances. The term *kinetics* deals with both the molecular pathways and the rates with which chemistry takes place. Chemists have known for decades that chemical combinations and rearrangements commonly occur very swiftly. Only in recent years have they been able to see just how swiftly. Until these advances, which became possible with the invention of the laser, measurements of chemical reaction rates were tantalizingly slow—on the scale of millionths of seconds.

In recent years, however, remarkable changes have come about from the science of laser optics that permits generation of light pulses a millionfold shorter than the time scale previously accessible. These extremely short pulses make possible the interrogation of the energy exchanges actually taking place in a chemically reacting system. These observations occur in *real* time, not after the fact of the chemical events. They allow direct observation of the dynamics of chemical processes. This science is still in an early stage, but its potential for the direct measurement of reaction rates in materials is very great.

Chemical Theory

Much of materials science, as shared with chemistry and metallurgy, has been largely empirical in character. Thus, much of synthetic chemistry is based on a very large body of laboratory experience that contains a strong predictive component but does not stand upon theories of atomic forces and energies. Yet we know that the complexities of chemistry must depend on the factors that govern simple atomic and molecular events. From mathematical calculations on these simplest systems there has evolved an extensive body of theoretical understanding of more complicated systems. The research has been tremendously helped by the development of high-speed, high-capacity computing machines. Now, with the aid of the science and technology of supercomputers, calculations can be made of the structures of a number of small polyatomic molecules with high precision. Rates of reaction of small molecules at surfaces can be studied. The chemical behavior of ensembles of molecules can be qualitatively simulated by calculation. Such representations are valuable, for example, in descriptive studies of atmospheric chemistry under environmental influences.

Similarly, interactions among molecules in water and in aqueous solutions have become well understood through theory. Here, the computer is asked to simulate the motions of large numbers of molecules as they travel and collide. The calculations can be extended to predict such properties as density at extremes of temperature and pressure, and thus provide information that would be hard to measure experimentally. Theoretical calculation is also successfully applied to study the ways that small molecules are attracted to the surfaces of crystalline solids and then react with one another. Such information is important in revealing what features of the molecular environment cause the reaction of atoms and molecules: with the aid of the computer it is possible to "see" molecules joining, coming apart, or moving over surfaces.

Purity

A classic part of chemistry has been the search for purity of reagents and products. The concept of the 19th century chemist was that progessive refinement of an element would yield a unique material, characterized by mass and other attributes such as the temperature of melting. Purity is a complex matter in most chemical compounds, because factors related to chemical architecture commonly come into play. Futhermore, the properties of even nominally pure elemental solids may be influenced by irregularities in the arrays of the atoms.

Nowhere are the issues of purity more important than in the making of electronic devices. In such uses the aim is for highly pure material (such as silicon) but with the deliberate addition of small amounts of "dopants" (such as arsenic and boron) that control the electrical properties of the material. This addition of deliberate impurities to a highly purified matrix is controlled with exquisite precision, in both quantity and position. All this work on materials for electronics is in fact the outcome of chemistry, and indeed the fabrication is a form of chemical synthesis.

Instrumentation

The advances of science have always depended upon the ability to observe events, and instrumentation is at the core of this ability. Certainly this the case in chemistry and materials science, where huge gains in the sophistication of intrumentation have occurred, in large measure from the successes of the sciences themselves. The gains are first measured in vast savings of labor and time, through major improvement in productivity in the gathering of information. However, far more importantly, major advances in instrumentation usually lead to unforeseen progress in understanding and to new ideas.

The health of the American economy in a competitive world depends on the successes of these interactions.

We have already noted the pervasive influences of computers and lasers. Much of the early discovery in instrumentation starts with physics, but then expands within chemistry, and becomes imbedded in materials science. Nuclear magnetic resonance (NMR) spectroscopy is a classic example of this trend. It came into being from studies by physicists of the magnetic character of certain atoms. Unexpectedly, information was also found that reflected the chemical environments that the atoms occupy. In the past two decades NMR spectroscopy has become a major instrument for chemists and is a standard tool in all modern laboratories. The continuing growth in the prowess of today's NMR spectroscopy depends upon new magnetic and electronic materials that were made possible through chemical research.

Perspectives

The coupling of chemistry and materials science is intrinsic to both disciplines. The health of the American economy in a competitive world depends on the successes of these interactions. The world will look more and more to these disciplines to improve the quality of life in the face of the ever-increasing needs of humankind.

Chapter 8

Simulation of Atoms and Molecules

William A. Goddard III

*F*or ages, artisans have relied on tradition, experience, and native skill to fashion utilitarian and decorative objects from metal, wood, clay, and other materials. Within limits, they are able to manipulate their ingredients to create attractive materials—brightly colored glasses, delicate ceramics, and flexible metal alloys—to suit various functions. Often relying on trial and error, however, they seldom understand in a fundamental way what gives different materials their particular properties.

Scientists are rapidly gaining the ability to design materials from the atom or molecule up. Starting with a list of desired properties, they can make educated guesses about what types of atoms or molecules, in what arrangement, would fit together to form the needed material. The new techniques for designing materials result from great strides made over the last few decades in the understanding of how atoms and molecules behave, starting with fundamental studies of the interactions between electrons and protons. Such studies yield insight into the forces that govern the dynamical behavior of atoms and molecules. This body of theory provides an atomic-level basis for understanding laboratory experiments and for making useful predictions about chemical behavior.

Computers in Materials Design

Computers have recently started to play a major role in materials design. By combining theoretical concepts with data from experiments, materials scientists and chemists can extract the information they need to simulate the behavior of a material. That step is accomplished by computing the dynamics of a collection of atoms or molecules. For example, working with an assemblage of

1473–5/88/0071 $06.00/0

5000 atoms represented by electrical signals in a computer, scientists can compute the system's temperature and other thermodynamic properties at any given time and track changes in the properties as time passes.

Computer simulations also help scientists penetrate hitherto mysterious processes, such as crystal growth. The production of large, single crystals of silicon and other substances has long been an art. The slightest disturbance—a jiggle, a cooling gust, an impurity, a speck of dust—can alter, sometimes unpredictably, the way atoms or molecules pile up to create a crystal. Such sensitivity tends to hide how a substance actually goes from a disordered fluid to an orderly lattice. By simulating the motion of individual atoms as they jostle and interact, researchers get information about the growth rate and the concentration of defects in a crystal.

Researchers can also gain chemical insights simply by watching on a computer display how a drug molecule actually "docks" at an enzyme. The observer, who may feel like a passenger on a weirdly shaped airplane arriving on a strange-looking runway, can get a sense of what the drug molecule sees when it arrives at a receptor site. Understanding the geometric shapes of receptor sites

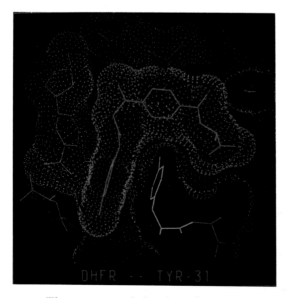

The structure of the chemotherapeutic agent methotrexate (blue dots) at the active site of the enzyme dihydrofolate reductase.

may eventually allow scientists to design drugs with fewer side effects. In place of a task that is like picking a lock to cure a disease, they'll be able to find a key perfectly suited to the lock.

Simulations become especially useful when materials designers can do experiments on a computer, without having to venture into the laboratory. Using interactive computer systems, they can add, subtract, or shift atoms, or substitute one atom for another, and then sit back and see what happens. If the outcome doesn't look right, they can try something else.

With fast computers and advanced graphics, interactive systems are available for a variety of applications. Such systems may become especially important for designing drugs, polymers, catalysts, and electronic materials. Using computers, scientists could design drugs to fit specific active sites of enzymes or genes. They could create new classes of polymers with specific chemical, electrical, or mechanical properties. Chemists could fine-tune alloy compositions and structures, and they could find additives to optimize the selectivity and yield of chemical processes, while keeping the quantity of undesirable side products to a minimum.

Preducting structure, from theoretical modeling, for bonding of a protein to the DNA binding site.

Designer Molecules

The driving force in the materials design revolution is chemistry: the molecular understanding of matter. What is fueling this revolution is the link that is being forged between individual atoms and actual materials. Starting with principles from quantum mechanics, scientists have made fantastic progress toward understanding the behavior and structure of molecules. They can now determine the stability of newly designed molecules even before the molecules are synthesized. For a specific chemical reaction, chemists can predict which bonds will break and what new ones will form. They can take a stab at predicting how quickly a reaction is likely to proceed, and they can try to determine what reaction conditions will minimize the yield of unwanted products while maximizing the yield of desired products.

To bridge the gap between molecular structure and the visible structure of materials, scientists are beginning to work out how atoms and molecules form into clusters or grow step by step into crystals. They are starting to see how clusters and crystals gather into grains. Grains, which make up the bulk of solid matter, are the features most likely to be visible to the naked eye or under a microscope. Scientists are also starting to understand how the properties of a chunk of metal, a shard of glass, or a lump of clay depend not only on the constituent atoms, but also on the material's macroscopic structure and on the cracks and other imperfections scattered within its bulk.

For instance, a polymer, at the molecular level, is a long chain of repeating units composed mainly of carbon, hydrogen, oxygen, and nitrogen atoms. That chain may be twisted into a spiral or have some other convoluted form. Under the right conditions, forces between different parts of a chain may cause the polymer molecule to fold into a neat package, or individual strands may interact and settle into orderly arrays in a single crystal. The properties of a polymer sample depend in some way on what happens at all scales. New chemical knowledge provides a bridge between invisible atoms and the plastics seen everywhere in today's world.

The movement back and forth between the microscopic level of work at a computer and the macroscopic design and implementation of experiments is likely to lead to a whole new way of scientific experimentation.

To simulate a material realistically, designers could start off with swarms of atoms or molecules interacting on a computer screen. Using data from these atomic-level simulations, they could gradually build up, level by level, a picture of the material as it appears in everyday life. It would be somewhat like starting with a microscope at its highest magnification, then decreasing the magnification so that more and more details are blurred. Eventually, only the overall nature of the material is apparent.

Such a step-by-step approach is clearly too time-consuming, even on the fastest computers, to allow designers to build and test their materials within a reasonable amount of time. Fortunately, increasing knowledge about the collective behavior of molecules and crystals now allows designers to start in the middle. They can speed up the design process by omitting some stages. Not all simulations have to go back to the atomic level. Judging by progress to date, designers may soon be able to use a computer to simulate a realistic system on a distance and time scale that matches real-world materials.

Laboratory results provide the checks that keep computer simulations from racing off into worlds far removed from reality. Measurable parameters that come out of computer models ought to match experimental data. If a designer, simulating a system and experimenting with the model, finds that the computer's predictions don't fit laboratory data, then it is back to the drawing board. Chances are that something important, perhaps the effect of some tiny impurity, has been left out of the model. Before the model can be safely used, the designer must find the factor that accounts for the discrepancy. That necessity may mean doing more careful laboratory experiments, or modifying the model, or both.

The movement back and forth between the microscopic level of work at a computer and the macroscopic design and implementation of experiments is likely to lead to a whole new way of scientific experimentation. In place of theorists and experimentalists doing separate jobs, one person would try both. One scientist could design a chemical system based on computer simulations and also perform laboratory experiments to check the simulation results and modify the theory. Many researchers find such a prospect tremendously exciting.

Fine-Tuning Catalysts

New design techniques are already being used to tinker with catalysts, which are materials that initiate, speed up, or slow down a chemical reaction. Typically, a catalyst does its job over and over again while itself remaining free of any permanent chemical changes. About one-sixth of the U.S. gross national product now requires catalytic manufacturing processes. Catalysts are especially important for refining petroleum into gasoline and other chemicals. They also play key roles in the manufacture of structural polymers and many of the newer materials used for integrated-circuit chips. Catalysts are also widely used in the production of synthetic fuels, for the direct conversion of methane to gasoline and other chemicals, and for producing hydrogen.

In the hottest field of catalyst research—the search for new catalysts for processing petroleum—molecular engineering has practically replaced traditional research methods. So far, much of that molecular engineering has focused on zeolites. Originally, zeolites were porous aluminosilicate minerals found in volcanic rocks. Now, dozens of synthetic zeolites, many of which have no natural counterparts, have been produced in laboratories and manufactured for use in industry.

Zeolites have a crystalline structure that forms a three-dimensional network of molecule-sized channels, which makes them ideal as molecular sieves. Like microscopic strainers, zeolites can separate the many ingredients of petroleum from one another. They also offer large surface areas where molecules can interact with elements in the zeolite; such interactions cause the molecules to crack into useful by-products. One main goal of researchers is to create new zeolites that crack larger numbers of molecules more efficiently, thereby increasing the amount of gasoline produced.

One type of synthetic zeolite known as ZSM-5 is used as a catalyst for converting methanol to gasoline. It is also helpful for upgrading gasoline octane numbers, an important consideration now that lead additives can no longer be used for the same purpose. ZSM-5 also has demonstrated a pronounced tolerance for the presence of impurities, such as metals, in crude oil and its derivatives. Commercial tests have shown that, unlike other catalysts, which can easily be poisoned by impurities, ZSM-5 survives the presence of significant quantities of nickel, vanadium, and sodium.

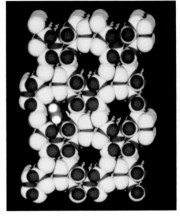

Zeolite ZSM-5 catalyst, used to convert methanol to high-octane gasoline.

Computer simulations help researchers visualize how different hydrocarbon molecules interact with a given zeolite, which allows some molecules to pass through its channels while keeping others out. Almost all of the accessible zeolite surface is in the interior of the structure. The general belief is that the selectivity of zeolite catalysts for making high-octane gasoline has to do with the fact that only certain molecules can move along the channels. There is a kind of molecular traffic control. The pore structure of the zeolite controls the approach of reactant molecules to the active site and the departure of products from it. Reactant and product molecules come and go along distinct but connected pore systems. Moreover, the geometry of sites at which reactions occur imposes constraints on the kinds of reactions that can take place and on the types of products that form. Of several closely related forms that may be produced, often only one particular configuration is favored.

One participant in this game of molecular hide-and-seek is the compound xylene. It consists of a six-membered ring of carbon atoms (a benzene ring), to

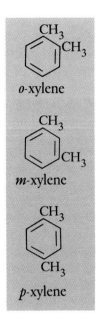

o-xylene

m-xylene

p-xylene

which two molecular groups (methyl groups) are attached. Because the ring has the form of a hexagon, the two molecular groups may be attached to adjacent carbon atoms or to carbon atoms at opposite vertices, or they may be attached to two carbon atoms separated by a single carbon atom with no attached group. It turns out that one of these three possible forms, or isomers, diffuses through a zeolite network a thousand times faster than the other two. Production of the fast isomer is strongly favored, because once it is created, it escapes much more rapidly than the others. The same site can be used to produce more of the fast isomer.

In a computer simulation, it is possible to watch xylene and other molecules moving along the channels in the zeolite. From such images, researchers can determine the optimum pathways for any particular molecule. In some places the molecules fits through easily, whereas in others it barely scrapes by. Computer simulations suggest ways of tailoring zeolitic structures to enhance the passage of molecules that participate in useful reactions.

The impressive zeolitic conversion of methanol to gasoline has yet to excite great economic interest, except in certain, very special circumstances. It would, however, have exciting economic consequences if scientists were able to produce methanol from methane. Thus, a major unmet challenge for catalytic chemistry is the one-step oxidation of methane to methanol. If a catalyst were developed for that purpose, it might lead to the production of synthetic gasoline by methane dissociation.

Zeolites are not the only catalysts worthy of intense study. Many economically significant reactions depend on the use of finely divided platinum as a catalyst. Platinum, however, is extremely expensive, and chemical engineers are interested in finding less costly alternatives. One application for platinum substitutes would be in catalytic converters that help get rid of nitrogen oxides and other pollutants in automobile exhaust.

Although computer simulations are a great new aid in the designing of catalysts, researchers still need to know more about how catalysis occurs. Many of the complicated processes that go on in catalysis are not well understood. Even where scientists understand the overall chemistry of which molecules come and go, they don't really understand in most catalytic processes what the catalytic surface itself actually does to promote the reaction.

As designers improve their ability to simulate chemical systems, they can begin to envision having computerized tools and models that track several processes taking place at the same time: the breaking apart of various materials, the chemical rearrangements that take place at surfaces, and the production of by-products from the reaction. No longer willing to tolerate even small amounts of toxic side products in industrial processes, scientists are trying to design specific catalysts for very selective chemistry. With on-screen simulations, chemists would begin to understand the nature of useful by-products, as well as those that just get in the way. They would be able to see how processes depend on surface features, such as site geometry and grain boundaries. Researchers are just getting to the point where they can begin using computers to explore some of the possibilities offered by theory.

Inside Superconductors

Some of today's most exciting chemical research concerns new, high-temperature superconductors. These ceramic superconductors, first identified in 1986, lose all resistance to electric current at temperatures as high as 90 K, well above the boiling point of liquid nitrogen but still far colder than room temperature. The standard way to make them is to mix and grind together the oxides or carbonates of their constituent metals. In the case of the

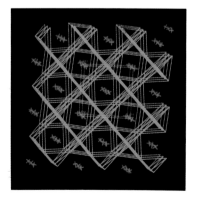

A distorted perovskite structure. Perovskites are some of the new high-temperature superconductors.

yttrium–barium–copper oxide, the material is made from a mixture of yttrium oxide, barium carbonate, and copper oxide. After heating, the powder mixture is pressed into a pellet and sintered at temperatures of around 950 °C. Later it is annealed in an oxygen atmosphere at a lower temperature (about 500 °C) to boost its oxygen content. The final step is slow cooling. The resulting material consists of a jumble of small grains sintered together, with tiny impurities trapped at the boundaries between the grains.

In the beginning, everyone was very impressed by how easy it was to make superconductive ceramics. Even high school science students prepared samples of yttrium–barium–copper oxide and demonstrated the compound's superconductivity by levitating a magnet over the samples, using easy-to-get liquid nitrogen to cool the material to sufficiently low temperatures. But making ceramic superconducting materials good enough for commercial applications takes a lot more hard work.

Chemists are now seeking better synthetic routes to ceramic materials and a more detailed understanding of their complex microstructure. It is clear now that a ceramic's microstructure and the conditions used to process it have a strong influence on the ceramic's electrical properties. By elucidating these influences, researchers hope to learn how to make superconductors with higher current capacity and other improved properties.

One of the factors that appears most crucial to the superconductive properties of any ceramic is its oxygen content, which depends on processing conditions. Oxygen content in turn determines the critical temperature at which the material first turns into a superconductor. At temperatures higher than the critical temperature, it behaves like an ordinary insulator or semiconductor.

Yttrium–barium–copper oxide has a unique structure, consisting of copper and oxygen atoms in the form of warped two-dimensional sheets and in linear chains of copper and oxygen atoms. These are interspersed between layers of yttrium or barium atoms. To get the best possible superconductive properties, the oxygen chains have to be very well ordered. And if the chain's copper atoms are replaced by other metals, such as aluminum or iron, the superconductive properties are degraded or lost. Heating up the material in an oxygen atmosphere drives

A SmCo$_5$ magnet being levitated above the bowl of the superconductor Ba$_2$YCu$_3$O$_7$ which is being cooled with liquid nitrogen.

some of the oxygen atoms out of the chains but leaves the planes intact. Eventually, the chains are disrupted enough so that the critical temperature begins to fall. The modified material must be cooled to a lower temperature than the original material before it can become a super-conductor.

The discovery of high-temperature ceramic super-conductors has fundamentally challenged previous theories designed to explain superconductivity. Most theorists now believe that the superconductivity seen at 90 K is different from the superconductivity that holds sway among materials such as niobium–tin alloys, which show the effect only at liquid helium temperatures.

The theoretical chemistry community is particularly interested in the properties of electrons in the new, ceramic superconductors. A few oxygen atoms at specific locations seem to control the hopping of electrons, which creates the electronic interactions that lead to supercon-ductivity.

To gain insights into the behavior of the new superconductors, theorists are putting their ideas through computerized tests. Using computer models, they try to make predictions, based on fundamental theories, about the structures and interactions within the new materials. The process has already led to the design and fabrication of new, related materials that are somewhat more stable than the high-temperature superconductors originally discovered. From recent studies of thin films of supercon-ducting materials, researchers can actually look at the effect film thicknesses have on the transition temperature. Theory, working together with experiment, is showing the way.

Mixing and Matching

The opportunities for molecular-level design of polymers and composites are nearly boundless. With the recent development of techniques for catalyzing the construc-tion of particular sequences of molecular links, one can almost think of building a polymer step by step. New polymers and recent advances in synthetic approaches promise to make feasible the selective synthesis of particular chemical and geometric configurations. Possibly, in the future, designers will create a material on the basis of the particular sequence desired and how it should be processed to get all the necessary ingredients in their proper places. Designers will determine what materials to make and what properties those materials should have. To do so, they need to know something about the properties of all the different configurations that polymer strands can take on when they are folded or when they interact with their neighbors.

As a result, polymer designers would like to use computers to answer a number of questions:

- How are relationships between stress and strain are affected by the structure of polymers?
- How do stress–strain relationships vary with time?
- How are stress–strain relationships coupled with activities at the molecular level?
- How do the strands interconnect?
- How well do the strands slip with respect to each other?
- How is slippage affected by the presence of any small bits of the starting materials left over from the reaction to create the polymer?
- What are the polymer's electrical properties?
- What are its optical properties?

Such considerations are important in the design of polymers to meet specific requirements.

All of those questions can be answered right now. Scientists merely need to put together the theory that is already in place to be able to make predictions on the right time scale for the designer to use a computer model based on the theory as a real tool for building polymers. Computerized polymer design is right on the horizon.

Late in 1987, a group of university and industry scientists, seeing the need for powerful simulation tools, organized a consortium to catalyze the development of suitable tools. Their aim was to stimulate development of research tools that would reduce the time needed to develop and test new polymer materials. Their first step was to develop a consensus concerning the features required in simulation and design tools for polymers.

Polymer Folding

In the study of polymers (as well as biopolymers, such as proteins), the basic problem is understanding how polymer strands fold. Polymers typically have strong interactions within polymer strands, but they often have weak interactions between the strands. The strong and weak interactions decide how a particular polymer folds up. The question polymer designers would like to answer is whether it is possible to predict the conformation that a given polymer is likely to slip into.

With the aid of a computer, the rational way to determine how a polymer is likely to fold would be to calculate the system's dynamics, taking into account all the forces involved. One would compute the movement of either a single strand or a collection of strands until all are settled in an equilibrium state. That step would mean applying Newton's laws of motion to thousands of individual atoms. The computations would continue until the system evolves to its proper structure.

What makes the rational approach impossible is that although a real polymer may take only seconds to rearrange itself and find its optimal configuration, a computer would probably have to take more than a million times a billion steps to reach the same answer. Even the fastest computers would take centuries to accomplish the feat. All the computers in the world today probably could not take a chemically interesting problem and come up with a proper answer. It would take too long.

Polymer designers are instead forced to take an irrational approach. Instead of passing logically from one conformation to the next until the optimal one is reached, designers can look at what the initial and final configurations ought to be and not worry about the details of how to get from one to the other.

To understand the contrast between the two approaches, imagine a man with his arms crossed on his chest. The man has his left arm on top of the right, and he wants to put his right arm on top of the left. The rational, dynamics approach would be to follow his arms' motion as they unfold, change their positions, and then refold. The irrational approach would be to let his arms cross right through each other so that their positions are reversed. There is really no need to worry about the fantastic energies or crazy conformations required for

intermediate steps, so long as the final position is satis-
factory.

One way to implement an irrational approach is to
allow an element of randomness. Imagine that fists and
elbows can move independently. Start them in one
position, shift them around randomly, reconnect them,
and see how close the system is to an optimal configuration. The only criterion is whether the end justifies the
means. In simulating physical systems, such a random
approach is known as a *Monte Carlo method*. It allows the
researcher to sample a large number of potential configurations very rapidly, whatever the starting point, and to
do it many orders of magnitude more quickly than if
rational forces led the way.

Monte Carlo methods can be used to simulate
processes such as a change of state, in which a material
goes from liquid to solid when the temperature goes
below a certain value. The idea is to start with a polymer,
say, a small one with just 128 carbon atoms and 250 or
so hydrogen atoms, at some high temperature. In
response to the energy available, the string of atoms is
allowed to wriggle about randomly. The computer
calculates the string's energy for each possible conformation, searching for a pattern that has the least possible
strain at that temperature. Eventually, the polymer strand
settles into an optimal configuration. The polymer is then
said to be in equilibrium. A similar computation can be
done for an interacting group of polymer strands.

Now the temperature is lowered step by step. When
the temperature gets low enough, the weak interactions
between polymer strands abruptly become important
enough to affect the motion of individual strands. The
strands settle into a definite pattern, and the system is
said to undergo a phase transition. Even though the
interactions between chains are weak and every bend
frozen into a polymer strand costs energy, there are
enough interactions to lock polymer strands into place. In
this way, a computer can simulate the way a polymer
condenses into a complex, folded structure. In the case of
polyethylene, which undergoes very weak folding, chemists can use such simulations to predict correctly that a
phase transition will occur, and to start looking at how
the folding occurs.

Computer-Aided Design

The whole approach of computer-aided molecular design,

or simulation, has great potential for solving a number of very important problems in materials science, especially when simulation works hand in hand with experimentation. The idea is to build useful models for complex systems, on the basis of results from atomic and molecular theory, but at such a level that designers can ask questions and get answers quickly. This kind of simplification—in which chemists no longer have to go back to first principles to compute realistic models of what happens in a material—is going to lead to a whole new way of chemical experimentation.

The model-building approach starts at the molecular level. Then it broadens from what is commonly thought of as chemistry—molecular interactions—into the macroscopic structure and composition of materials. Ceramics has already become chemistry. Instead of manipulating a mixture of oxides—heating them, breaking them down, or putting them together—to create a ceramic with certain properties, researchers can now start at the molecular level and engineer ceramic materials. They use sophisticated chemical techniques to construct new kinds of materials, achieving an unprecedented level of control over a ceramic's properties. As scientists learn more about how metal atoms are put together and use chemical techniques to create alloys, metallurgy too may follow the course of ceramics and become chemistry. The computer plays a key role in bringing chemistry into materials design.

Chapter 9

Materials for Advanced Electronic Devices

George M. Whitesides

*T*he typical integrated-circuit chip looks like an armored insect. It has a hard plastic or ceramic case from which an array of metal legs protrude. Inside this arthropodal packaging sits a remarkable collection of materials that are carefully engineered to control the movement of electrons.

The Structure of an Integrated-Circuit Chip

The key working parts of a generic chip are really too small to see with the naked eye. Only the packaging is readily apparent, but under a scanning electron microscope, you would see that the chip itself is made of a series of thin layers, with one material coated on top of another (*see* Figure 1). Each layer has a thickness that may be anywhere from a few atoms to several thousand atoms deep, depending on its function. These layers, carefully laid onto a chip's surface, then etched into vast arrays of microscopic electronic switches and gates, work together to shuffle electrons about. In place of the maze of separate wires and components typically associated with electrical circuits, the chip's wires and electronic devices are integrated as lines and channels on its surface. Today's integrated-circuit chips may carry hundreds of thousands of transistors, each measuring as little as a few micrometers across.

A typical chip starts off as a clean, polished wafer of silicon doped (deliberately contaminated) with a trace of either boron or phosphorus. Doping with phosphorus produces an *n-doped semiconductor*, which provides electrons as current carriers. Boron produces a *p-doped semiconductor*, which provides positive, electron-deficient

1473-5/88/0085$06.00/0

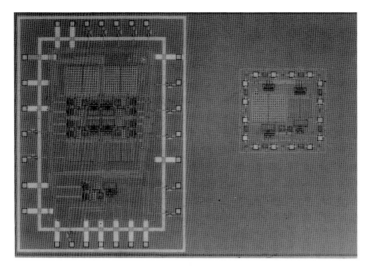

Figure 1. Scanning electron micr scopic view of an integrated-circui chip.

regions, called holes, as current carriers. Atop this silicon surface, chip fabricators deposit as many as a dozen layers, some of which, like silicon, are semiconductors, while others may be electrical insulators or conductors. Each type of material plays a role in pushing electrical charge through the chip's circuits. The conductor in an integrated-circuit chip may be a metal, such as aluminum, or silicon that is heavily doped with a conducting material. Strips of conductors form electrical connections within and between circuit elements. The insulator usually consists of silicon dioxide and is used to protect the silicon surface and separate conducting regions where no connections are desired. Figure 2 shows an integrated-chip cross section.

In the fabrication of integrated circuits, the layers are

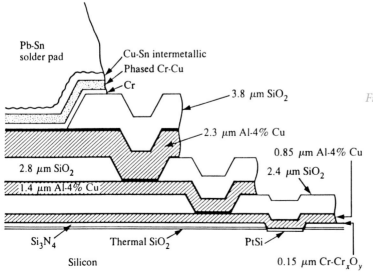

Figure 2. Cross section of an integrated-circuit chip.

added one at a time. A template called a "mask" determines the pattern for each layer. Fabrication takes place
by various steps that combine oxidation, mask protection,
etching, diffusion or ion implantation, and vapor deposition. Finally, the chips are sealed in a protective plastic or
ceramic package. The package is what many people
envision when they think of a computer chip, but the chip
itself is actually a flat piece of silicon, no bigger than a
fingernail.

These tiny chips, carrying an increasingly heavy load
of transistors and other devices, have fueled the explosive
growth in the electronics and computer industries. Every
year, more and more everyday products, from toys and
tape recorders to refrigerators and automobiles, rely in
some way on integrated-circuit chips.

Competitive Choices

Microelectronic technology has evolved so quickly and
has become the focus of such keen international compe-
tition that policymakers are now left wondering where
the United States should focus its efforts and what role
materials science should play in this research. On the
commercial side, a very large fraction of the gross
national product has come to depend, more or less
directly, on electronics. More recently, from the perspective of national security, there is a growing perception
and concern that many essential electronic systems in
military applications use components that are available
only from Japan.

The growth of microelectronics raises two important
issues. First is the basic question of how best to try to
defend the existing electronics industry, which relies
heavily on silicon-based technology. Silicon itself will
remain an important electronics material for the foresee-
able future, especially as researchers achieve smaller
feature sizes and three-dimensional structures. New
technologies based on gallium arsenide and the construction of hybrid devices that combine the most promising
characteristics of silicon and gallium arsenide will find
their places. Technologies based on diamond or cubic
boron nitride are also possible but much more distant.
New superconducting materials will probably have a role
in high-speed circuit connections.

The second issue is to determine in what directions the technology should be pushed. The electronics industry is continuing to move in the direction of miniaturization. Features on silicon chips are becoming so small that they are easily dwarfed by a human hair, and miniaturization can be carried further still. To achieve the next level of ultra-large-scale integration, chip designers are venturing into the realm of three-dimensional structures and exploring exotic approaches such as layered superlattices. New techniques of fabrication promise unprecedented densities and device geometries. The new generation of chips would have features that are less than a micrometer across; millions of transistors would be packed onto a scrap of silicon or some other semiconductor material. By studying the complete range of properties of electrons in all types of materials, scientists provide the fundamental knowledge essential for large-volume, reliable production of microelectronic devices.

As chip designers advance to the stage of ultra-large-scale integration, they will try at the molecular level to design the properties of chips. Then it will be especially important to be able to predict the response of a particular combination of materials and a particular circuit design. Chip designers will also need the ability to assess various strategies for growing the materials required for the microscopic structures they desire.

Packaging Semiconductors

The plastic or ceramic packaging that contains a chip's links with the outside world and protects it from damage is now becoming the bottleneck that restricts efforts to increase the speed and shrink the size of integrated-circuit chips. Although packaging has traditionally been much less exciting to work on than the chips themselves, new developments in packaging may be the key to breakthroughs in chip technology.

Most electronic devices are placed on alumina (aluminum oxide) supports, or substrates. A piece of alumina is made by compressing alumina powder into an appropriate form, and then sintering (heating without melting) it to create a dense microstructure (*see* Figure 3). There are important correlations between the form of the particulate alumina powder and the electrical properties

Figure 3. Sintering of alumina. left, ideal packing; right, dense microstructure obtained at 1300 °C for one hour.

and dimensional stability of the final assembled product. One important goal is to perfect the final product by using the best available alumina powder, such as the highly regarded material provided by the Japanese company Sumitomo. That kind of perfection is likely to lead to fundamentally new processing technologies. Some researchers are also trying to fabricate more complicated alumina structures that not only support the functions in an integrated circuit, but also perform important functions of their own.

Atomic Views

To control the materials and the chemical reactions involved in fabricating sophisticated semiconductor circuits, scientists need the ability to detect surface details as small as individual atoms. To achieve this notoriously difficult task, chemists are beginning to adopt techniques developed in the field of vacuum physics. The scanning tunneling microscope enables researchers for the first time to image atoms directly, one at a time.

In a tunneling microscope, an extremely sharp, metal needle is brought within a few angstroms of the sample's surface. This distance is small enough for electrons to leak or tunnel across the gap and generate a minute current. As the gap between the tip and the sample increases, the current decreases. As the probe crosses the sample, moving back and forth across its surface, its vertical height is continually being adjusted to keep the current constant. In essence, the probe traces out a contour map of the sample's surface atoms.

The invention of the scanning tunneling microscope and subsequent refinements in its design have given scientists increasingly sharp views of atoms perched on solid surfaces. Most recently, the microscope has provided pictures not only of silicon atoms neatly arrayed on a silicon surface but also of the bonds holding the atoms in place (*see* Figure 4). Normally, the voltage applied between the sample and probe stays the same. To observe the bonds between atoms, scientists at the IBM Thomas J. Watson Research Center (*1*) held the probe still while varying the voltage. The result was a map of how the current varies at selected points over a surface. The information was then used to show where electrons bonded to surface atoms were likely to be.

It's amazing that scientists can make such minute observations and can begin to use devices like the probe to pick up individual atoms and move them to some other part of the substrate. Such a feat was inconceivable 20 years ago.

Speed and Gallium Arsenide

Since the early 1970s, scientists have been promoting gallium arsenide as a faster, more efficient substrate material than silicon for making integrated-circuit chips. (Figure 5 is a scanning tunneling micrograph of gallium arsenide.) However, the vast majority of chips are still made from silicon, which is abundant and cheap. The most important advantage of gallium arsenide is speed. Electrons travel about five times faster in gallium arsenide than they do in silicon. Gallium arsenide also has a high resistance to electrical current before it is doped with any impurities to form circuit elements. Consequently, a gallium arsenide wafer, or substrate, is semi-insulating, whereas a silicon wafer is semiconducting. That feature simplifies gallium arsenide circuit fabrication considerably. Gallium arsenide also offers a wider range of operating temperatures than silicon and much higher radiation hardness, which is a decisive advantage for military and space programs. Another major advantage is that gallium arsenide can be doped in such a way that it emits light, which makes it useful for lasers and light-emitting diodes.

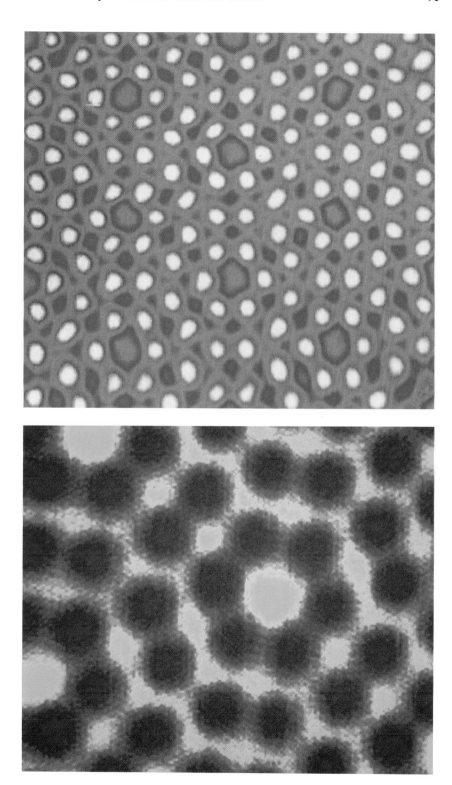

Figure 4. Two views of the silicon surface by scanning tunneling microscopy. The larger is a blow-up of the smaller.

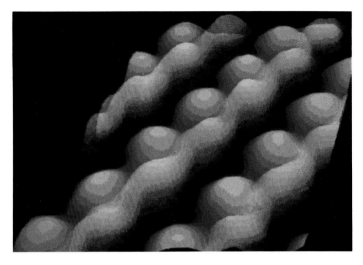

Figure 5. Scanning tunneling micrograph of gallium arsenide.

The problem with gallium arsenide is that the material is exceptionally difficult to grow into large, defect-free crystals. Much care is needed to produce from the elements gallium and arsenic a very precisely tailored compound with just the right properties and the right proportions. Large silicon crystals, on the other hand, are relatively easy to produce, in part because only one element needs to be controlled. With gallium arsenide, two materials must behave properly. One of those materials—arsenic—is toxic and volatile at the high temperatures needed to grow crystals. It tends to bubble out of the high-temperature melt. Despite the development of various methods for overcoming these problems, high-quality gallium arsenide is still relatively expensive and hard to get. Furthermore, silicon is a better heat conductor, and it allows more transistors and other devices to be packed into a given surface area.

A Match for Silicon

Until now, silicon and gallium arsenide technologies have developed somewhat independently. One way of dealing with the silicon/gallium arsenide trade off would be to marry the two types of components. Putting gallium arsenide semiconductor circuits atop a silicon base is a bit like mating a Ferrari with a Honda. The components seem incompatible, but if the match were to work, the result would be an attractive combination of high performance and economy.

Hybrid integrated-circuit chips of gallium arsenide and silicon may now be feasible. Researchers at the University of Illinois at Urbana–Champaign have discovered a way to deposit gallium arsenide layers on top of silicon wafers without spreading the crystal defects that ruin the electronic properties of the materials.

The trick is to find a way of aligning the silicon and gallium arsenide crystal lattices. Normally, the structures do not quite match. For a row of 25 silicon atoms, only 24 atoms from a gallium arsenide layer are needed to fill the same space. Aligning the two materials produces a large number of defects where the two lattices meet. The mismatch can be overcome if the silicon base is slightly tilted. A gentle slope of about 4° provides, at the atomic level, tiny steps that take care of the problem. If these steps have the right orientation with respect to the silicon crystal lattice, then the inherent bumpiness of the slope does not produce dislocations that thread their way into the gallium arsenide layer.

The orientation is the key. For a square silicon chip with an upper surface parallel to a face of the crystal lattice, the slope needs to rise from its low point at one corner to its peak at the diagonally opposite corner. The combination of light-emitting gallium arsenide chips and complex, tightly packed silicon circuits could make it possible to connect circuits optically instead of using wires. In some of today's most advanced chips, far more power already goes into driving the wires that connect chips than in running the complicated silicon circuits themselves. With hybrid chips, the wires connecting one device to another could be replaced by an efficient optical system, perhaps using optical fibers.

Because all parts of an integrated circuit need not be equally fast, it may eventually be possible to deposit gallium arsenide at only the points on a silicon circuit where the chip must operate quickly. Recent work at the University of Illinois (2) will probably accelerate the pace of hybrid-chip research. Continuous lasers and optical interconnects may be developed soon. More and more research groups are active in the field, and several small companies have been established to develop the technology.

The use of materials that respond to light suggests the potential for a major shift in technology from computing and communications devices based on the movement of electrons to devices based on the transmission of light. Optical computation and communications are not yet a major commercial technology, although the communications part is becoming important. One major interest of researchers is to develop classes of materials that enable one to manipulate light with the speeds and characteristics that are required of new generations of chip-to-chip and continent-to-continent communications. Perhaps future computers will do their computations by manipulating light pulses instead of electrons. There is much opportunity for important inventions and major new applications.

Layers upon Layers

One interesting type of microelectronic device now being studied is a heterojunction device made up of multiple, alternating layers of gallium arsenide and gallium aluminum arsenide (*see* Figure 6). Each layer is only a couple of atoms thick. With such small structures, a host of remarkable phenomena begin to emerge. One is negative resistivity: When the voltage is increased, the current goes down rather than up.

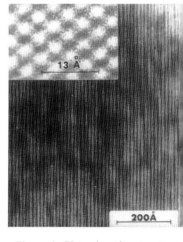

Figure 6. Heterojunction structure.

Another phenomenon evident in exceedingly thin semiconducting films is ballistic transport, which allows an electron to pass from one side of a barrier to another without striking any atoms in between, tunneling through the barrier like a ghost passing through a wall. Under normal conditions, electrons do not race from point to point like speeding bullets; they stumble along more like drunken sailors. Flowing through the circuitry of a chip, they constantly bump into impurities, rebound off walls, and slow down as they pass through the electronic gates that signify on or off in a microprocessor. Each collision costs distance and time. Ballistic transistors are designed to be so small that an electron can shoot right through the device with scarcely a single collision. The idea is to make the length of the region that electrons have to travel comparable to the average distance they go before colliding.

Over the last decade, there have been a number of spectacular advances in the construction of semiconductor heterostructures with a specified band gap. These have led to the discovery of the totally unexpected two-dimensional behavior of electrons, as well as to the development of novel electronic and optical devices. Many of the advances are a direct result of the development of new crystal growth techniques allowing the formation of layered semiconductors that are perfect on the atomic scale.

A number of technologies make possible these astonishing structures. One is metallo-organic vapor-phase deposition, also called vapor-phase epitaxy. It involves taking appropriately prepared organometallic compounds and allowing them to react in the vapor phase to deposit the desired inorganic structures on an appropriately prepared substrate. Epitaxy is the process of growing crystalline semiconductor films in which the substrate determines the crystallinity and orientation of the thin films grown on top of the substrate. Among the techniques that have been widely used are liquid-phase epitaxy, chemical-vapor deposition, and molecular-beam epitaxy.

In liquid-phase epitaxy, the epitaxial layer is grown by cooling a heated metallic solution saturated with the components needed to grow the layer, while that solution is in contact with the substrate. In chemical-vapor deposition, the epitaxial layer is grown from a heated stream of gaseous elements or compounds, which react at the surface of the substrate. Recently, researchers have made considerable progress in growing lasers and other quantum-well heterostructures, using metallo-organic chemical-vapor deposition for epitaxial growth.

The most advanced semiconductor heterostructures require special fabrication techniques under controlled, high-vacuum conditions. The process of molecular beam epitaxy is a bit similar to painting with spray guns containing different-colored paints. The materials to be layered are heated in separate ovens within a vacuum chamber until their atoms begin to boil off. A computer-controlled shutter then opens and closes at precisely timed intervals, releasing the proper quantity of atoms, first of one material then another, from each furnace. The atoms strike and adhere to the base plate, forming alternate layers.

Nevertheless, the precise control of the chemical reactions that take place at surfaces and especially the control of the purity of the materials is one goal that heterostructure fabricators have not yet attained. Whoever first learns how to achieve such control—whether it happens in Japan, the United States, or elsewhere—will dominate a substantial part of electronics processing in the future. The challenge shows a very real need for technological innovation.

Warmer Superconductors

For more than a decade, researchers have been toying with the idea of building integrated circuits for computers using superconducting Josephson junction switches. A *Josephson junction* consists of two thin layers of superconducting metals, such as lead or niobium, which act as electrodes, separated by an even thinner insulating layer. At liquid helium temperatures, the metal's electrical resistance drops to zero, and electron pairs can tunnel across the insulating junction. An externally applied voltage can stop the current flow, and thus this device can be used as a switch. Despite the disadvantage of having to work at temperatures close to absolute zero, Josephson junction circuits have appeared attractive for computer circuits because they switch on and off faster and emit only one one-thousandth as much heat as semiconductor transistors.

Until recently, however, researchers had had very little success in finding materials that become superconductors at higher temperatures. The best they could find were certain metal alloys that abruptly lose their electrical resistance at temperatures below 24 K. In 1987, the situation changed dramatically with the discovery of special ceramics that remain superconductors at temperatures now as high as 90 K. Because that temperature is greater than the boiling point of liquid nitrogen, though much lower than room temperature, much less costly refrigeration techniques can be used to cool the ceramic materials enough to turn them into superconductors. Therefore, the number of potential applications is greater.

At the moment, scientists are optimistic for two reasons. They know they can shift the critical temperature at which a material becomes superconducting by varying the composition and the structure of these new superconducting ceramics. They are also certain that investigations of how these materials achieve a superconducting state

. . . chemists and engineers have joined the exhilarating quest to understand high-temperature superconductors, improve their properties, and push them into practical commercial applications.

will lead to the elucidation of fundamentally new mechanisms for superconductivity. Both advances will likely lead to devices—from electronic circuits to electromagnets—that operate at relatively high temperatures but pose no resistance to electric currents. For the world that we live in, these advances are potentially as important as advances in molecular biology.

From a materials point of view, the greatest amount of attention has been focused on yttrium–barium–copper oxides. However, these are not the only ceramic materials that show superconductivity. Researchers are beginning to look at other possibilities, especially as they learn what special electronic properties to seek in specific materials. A host of chemists and engineers have joined the exhilarating quest to understand high-temperature superconductors, improve their properties, and push them into practical commercial applications. One early application may be for magnetic field detectors and simple electronic devices.

Diamond's Sparkling Potential

A fiery sparkle isn't all that makes a diamond so eye-catching. Its hardness and its ability to conduct heat and to act as an electrical insulator make diamond an attractive material for electronic circuits designed to survive high temperatures or withstand intense radiation.

Although it is hard to imagine a way to fabricate diamond into thin sheets of the sort used for silicon-based devices, some researchers believe that a future generation of electronic devices may be based on diamond—if they can overcome certain problems. What is needed is an economical, practical method for laying down and then etching thin diamond films on silicon and other surfaces.

Diamond is attractive because it carries electrical pulses extremely quickly. Its transparency means that it can transmit optical signals. Because diamond is the best known thermal conductor, it could be extremely efficient in diamond-based electronic devices. Complete impermeability to oxygen and similar species gives diamond many of the properties one can hope for in an almost ideal device.

The basic process for generating diamond coatings

involves passing a gaseous mixture of methane and hydrogen molecules at atmospheric pressure through a microwave bath. This process breaks up the molecules into hydrogen and carbon atoms, which can then settle onto a silicon surface. This chemical vapor deposition technique is not unlike that used for gallium arsenide structures.

The presence of hydrogen appears to be necessary to ensure that carbon atoms end up in a tetrahedral diamond crystal arrangement rather than in a planar graphite structure. Hydrogen atoms seem to pick up "dangling" bonds on a freshly laid carbon surface, which prevents the carbon's structure from collapsing into the form of graphite. Moments later, carbon atoms replace the hydrogen atoms, and the crystalline diamond film continues to grow.

It takes about an hour to lay down a 1-micrometer-thick diamond layer. Each film consists of a random array of individual diamond crystals about 200 angstroms across. Researchers are now trying to speed up the deposition rate and to build films that consist of a single diamond crystal. That accomplishment should make diamond fabrication very simple. The new process is potentially cheaper, cleaner and more versatile than high-temperature, high-pressure techniques now used to produce synthetic diamonds.

A diamond film's first application may be in microelectronics. Because diamond conducts heat like a metal, tiny diamond slabs could be used as bases for electronic circuits that need to survive high temperatures. Conventional silicon chips usually cannot withstand temperatures greater than 300 °C. However, diamond-based devices could be used as sensors in engines or nuclear reactors. Furthermore, because diamond does not overheat easily, more circuit elements could be packed onto a diamond-based chip than on a silicon chip.

In the United States, scientists at the Naval Research Laboratory in Washington, DC, and at MIT's Lincoln Laboratory have long worked on designs for diamond semiconductor circuits. Until recently, they lacked materials on which to test their designs. New research efforts to produce diamond films at Penn State; North Carolina State University in Raleigh; and at the Research Triangle Institute in Research Triangle Park, NC, will now provide the essential materials for that work (3).

Early in 1987, the Japanese company Sumitomo announced that it had succeeded in developing a diamond semiconductor. The diamond film is doped with a small amount of phosphorus, which turns it into an n-type semiconductor. The new process brings scientists one step closer to creating true diamond transistors and other electronic devices. Researchers working at the MIT Lincoln Laboratory (*4, 5*) have created authentic, though highly primitive, transistors in a thin diamond film by spraying ions in patterns in the presence of nitrogen dioxide trapping devices.

The Japanese company Mitsubishi has already come up with one commercial application for diamond in information systems. The company now manufactures a very fast Winchester drive in which the magnetic medium operates without a lubricant. With the head in very close proximity to the device, a diamond-film wear-resistant barrier prevents catastrophic crashes in the event of occasional, inadvertent contact between the head and the spinning disk.

Looking Ahead

Recent advances in high-temperature superconductivity and the fabrication of thin diamond films are two significant signposts pointing toward technologies that may someday play crucial roles in microelectronics. Microelectronic technologies are changing very rapidly, and any nation that expects to remain at the forefront of new technologies should invest broadly in investigating materials that have the potential for dramatically changing the world. The dividends may not come this year, or next year, but perhaps 25 years from now. Nevertheless, the investment is necessary to ensure a secure economic future.

References

1. Binnig, G.; Rohrer, H. *Angew. Chem., Int. Ed. Engl.* **1987,** *26,* 606.
2. Morkoc, H.; Fischer, R. Eur. patent application EP 232082 A2, 1987.
3. DeZries, R. C. *Annu. Rev. Mater. Sci.,* **1987,** *17,* 161.
4. Geiss, M. W.; Rothman, D. D.; Ehrlich, D. J.; Murphy, R. A.; Lindley, L. T., *EDL-8,* **1987,** *8,* 341.
5. Geiss, M. W.; Efremon, N. N.; Rothman, D. D. *J. Vacuum Sci. Technol., in press.*

Chapter 10

Chemistry of Materials for Energy Production, Conversion, and Storage

Mark S. Wrighton

From the time that our ancestors first domesticated fire, human control of energy has fueled the growth and progress of civilization. At first, the concept of energy as a source of heat and light was inextricably linked to the image of luminous flames licking scraps of wood or the hot glow of lumps of coal. Then, the availability of petroleum and the invention of the internal combustion engine initiated a vast transportation system and a new mobility. In recent times, scientists have found sophisticated ways to harness energy contained in the heart of radioactive atoms or in light from the sun. Each new energy technology extends the repertoire of human activities.

In 1985, a National Academy of Sciences committee, in its report *Opportunities in Chemistry* (*1*), commented, "As we look ahead there is no doubt that the nation's wealth and quality of life will be strongly linked to access to energy in large amounts." Continued access to large energy supplies now depends on the development of new sources of energy, on improved methods for extracting energy from old sources, and on making the most of any energy source available by putting a premium on efficiency. At the same time, governments and industries are being pushed toward energy technologies that are safer and generate fewer hazardous by-products than current technologies.

Both existing and newly emerging energy technologies would be impossible without the development of high-performance materials. And the development of such materials relies heavily on chemists, metallurgists, ceramicists, and others, who are responsible for their preparation and efficient manufacture. Future developments in the production, conversion, and storage of

1473–5/88/0101 $06.00/0

energy will depend on advances in basic chemistry and in materials and chemical processing.

Three chemical substances—coal, petroleum, and natural gas—account for the bulk of energy used in the world today, and they will continue to play an important role for decades to come. To use such energy-rich materials efficiently, scientists and engineers need to understand and control the chemical reactions underlying the production, refining, and combustion of fossil fuels. Although the depletion of fossil resources is inevitable, new opportunities in chemistry-related disciplines promise to increase the prospects for continued access to abundant energy supplies.

U.S. energy use and demand for imported oil are bound to continue increasing and further strain the trade deficit. Some projections even suggest that within just five years, oil imports alone will equal the present yearly trade deficit. Meeting our long-term energy needs will hinge on the development of new energy sources as well as more effective use of existing resources. Basic research in chemistry and chemical engineering could eventually offer new possibilities for reducing oil imports and correcting the trade imbalance.

Yesterday's basic research on energy-related materials and processes is now paying off handsomely. One excellent example of a successful practical application stemming from basic chemical research is a process called *heterogeneous catalysis*, the use of special substances that alter the rates of chemical reactions without themselves undergoing any permanent change. An estimated one-sixth of the value of all goods manufactured in the United States now involves heterogeneous catalysis and other catalytic processes.

Little is known about catalysis at the molecular level. Catalysts, such as the metal platinum in powdered form or synthetic aluminosilicate minerals known as zeolites, somehow provide congenial environments for chemical reactions. At surface sites on the catalysts, molecules meet, disassemble, and recombine into new, more useful forms. During the past two decades, scientists and engineers have developed a variety of techniques for studying, at the atomic level, the structure, composition, and chemical bonding of guest molecules at catalytic surfaces. Much more is still to be learned about catalytic materials and what happens during catalysis. Fundamental

Yesterday's basic research on energy-related materials and processes is now paying off handsomely.

advances now possible in the molecular-level understanding of catalysis promise to yield an important knowledge base for future energy technologies.

Material Boundaries

A recurring theme of energy technologies is that *interfaces*—the boundaries between different materials—are often a crucial aspect of a technology. In heterogeneous catalysis, for example, the key processes take place at interfaces between a solid and a reacting medium. Crucial to the functioning of solar cells are interfaces between p-doped and n-doped semiconductors, between metals and semiconductors, and between semiconductors and liquids. Corrosion in nuclear power plants can stem from interactions between thin-walled metal tubes and superheated water within steam generators. Efficient tertiary oil recovery depends on knowledge of how minerals trap oil. Interfaces are even crucial to the process of plant photosynthesis, nature's solar conversion system.

To understand the function of an interface, one must know its chemistry, that is, its composition and structure. Once chemists understand where molecules settle on the surfaces of catalysts and how they rearrange themselves, they will be well on the way toward controlling catalysis. Current research focuses on such questions as:

- Which areas on the surface of a catalyst are the most important for catalysis?
- How long do molecules remain on the surface, and does this residence time affect the efficiency of catalysis?
- What sorts of chemical intermediates give the optimal selection of products?
- What are the mechanisms by which catalysts become spoiled, or poisoned?
- Why are transition metals such as platinum best suited as catalysts?

Historically, chemists have made progress in understanding chemical reactions by learning about composition and structure at the atomic or molecular level. Reactions of molecules, large and small, natural and synthetic, have been studied through the application of X-ray diffraction to determine structures of crystals.

Where crystals have been difficult or impossible to obtain, researchers have used other methods, often involving spectroscopic techniques of structure solving. By looking at the interaction of matter with electromagnetic radiation, whether radio waves, visible light, or X-rays, researchers have been able to make educated guesses about the nature of the atoms and molecules that affect the incoming radiation.

Over the past few decades, a battery of spectroscopic techniques using photons, electrons, ions, atoms, and molecules as surface probes has emerged to answer questions about molecular structure. One such tool, *Auger photoelectron spectroscopy,* provides unique information about the composition and chemical state of surface materials.

In Auger spectroscopy, electrons can be used to knock out other electrons from deep within an atom. Once an electron has been expelled, the remaining electrons in the atom rearrange themselves. During the rearrangement, another electron may be lost, in a phenomenon called the Auger effect. The energy and intensity of the lost, or Auger, electrons identify the atoms that emitted them, and elucidate the nature of their chemical environment.

Researchers can use techniques like Auger spectroscopy in a single research-group laboratory setting, which is typical of a major research center. In addition, national facilities provide access to knowledge about interfaces. These include the Brookhaven National Laboratory, Stanford University, and other institutions with *synchrotron light sources,* huge machines that whirl electrons around at speeds close to the speed of light. The whirling electrons, in turn, emit concentrated, directed bursts of X-rays. (*See* Figure 1.)

A few years ago, a team of Stanford researchers used their synchrotron to study an industrial catalyst consisting of cobalt, molybdenum, and sulfur. This particular catalyst removes sulfur—a substance that poisons certain other catalysts used in petroleum refining—from crude oil in a process called hydrodesulfurization. During the reaction, the researchers measured the way the surface of the catalyst absorbed X-rays over a range of energies. They used this information to calculate the distance between atoms at the surface of the catalyst and to determine how various atoms at the surface pack around

Figure 1. General view of the experimental floor of the vacuum ultraviolet ring of the national synchrotron light source at Brookhaven National Laboratory. This is America's largest facility dedicated solely to the production of synchrotron radiation. It provides intense beams of X-rays and ultraviolet light for use by scientists from all over the country for research in physics, chemistry, biology, and various technologies. (Courtesy of Brookhaven National Laboratory.)

each other. By systematically altering the composition of the catalyst, the team correlated the changes in the activity of the catalyst with changes in surface structure. In that way, scientists and engineers hope to find the active sites, or regions on the surface where catalysis is thought to occur.

Recent developments in microscopy also make it possible to establish the structure of interfaces—particularly interfaces involving solid materials—at unprecedented levels of detail. A new tool, the *scanning tunneling microscope,* provides atomic-level structural resolution under a wide range of conditions and for a wide variety of materials. It is already being applied to semiconductor interfaces for use in solar devices and microelectronics, and to catalyst surfaces that are important in energy conversion systems. Such information has been vital for understanding molecular chemistry and is likely to be crucial to the development of rational design and fabrication of new interfacial material systems tailored for specific purposes.

Together, the methods and instruments available to characterize interfaces comprise a powerful arsenal that will allow scientists to attack increasingly complex problems in interface science. Those intellectual challenges are drawing the attention of some of the nation's best scientists and engineers. Armed with new tools and methods, they are taking on challenging problems and already making significant strides in some areas. As researchers explore and begin to exploit novel combinations of materials, interface science is likely to have an enormous economic impact on many technologies, including microelectronics, high-strength materials, and energy.

Chemical Electricity

One promising device for energy production is the fuel cell. *Fuel cells* are devices that convert chemical energy, or fuel, directly into electrical energy. They work on the same principle as batteries. The main difference is that the cells are continuously fed with fuel, and they can be used as electricity. Many advantages are associated with converting fuel directly to electricity rather than burning the fuel as we do now in fossil fuel-fired electric power plants. Fuel cells help minimize noise and air pollution. They have a high theoretical efficiency, and they require no moving parts. Fuel cells have been used to generate electricity in manned spacecraft (*see* Figure 2) and other specialized applications, but much more research is needed to make fuel cells efficient for everyday use in power plants and even in automobiles.

Another device that produces electricity is the *photoelectrochemical cell*. Essentially a semiconductor immersed in chemical soup, it can convert sunlight

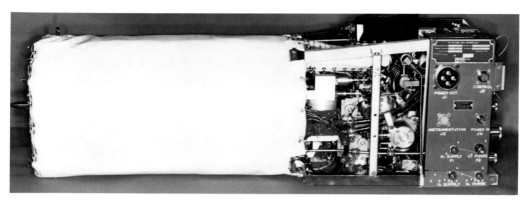

Figure 2. The space shuttle Orbiter fuel cell. The shuttle is equipped with three such fuel cells that produce electrical power by the electrochemical conversion of hydrogen and oxygen. (Courtesy of International Fuel Cells.)

directly to electrical or chemical energy, or to a combination of the two. The basic device consists of two electrodes that are connected by an external circuit and immersed in an electrolyte solution. One or both of the electrodes is illuminated with sunlight. The interaction of light with the material induces a flow of electric current in the external circuit, as well as chemical reactions at the interface between each electrode and the electrolyte solution.

The efficiency and effectiveness of fuel cells and photoelectrochemical devices depend on the interfaces where chemical reactions occur. In both fuel cells and photoelectrochemical devices, the reactions that take place at interfaces, called interfacial catalysis, are crucial to energy conversion.

Recent progress in the study of interfacial electron-transfer catalysis is at the core of some exciting developments in the chemistry of energy-related materials. In fact, such progress is the key to future developments in energy conversion technology. Tailoring catalytic interfaces for electron transfer also has other practical consequences. Currently, about seven percent of all electricity generated in the United States is used for synthesizing electrochemical products, such as chlorine and aluminum. Improving the efficiency of electrochemical processes would do much to help conserve energy and lower the cost of such synthetic products. No matter what energy technology is used for generating electricity in the future, advances in interfacial catalysis will be important for conserving energy.

Improving Fuel Cells

The most common fuel cell is one that combines hydrogen gas and oxygen gas to generate electricity and water. Often a fuel processor unit first converts natural gas or another hydrogen-containing fuel into a hydrogen-rich mixture. In the cell, that mixture is fed to one electrode, while oxygen (from the air) is supplied to the other electrode. Between the two electrodes is the electrolyte, such as phosphoric acid. The electrolyte functions much like the sulfuric acid in an automobile battery and permits an electrochemical reaction to take place. Hydrogen molecules, filtering through their electrode, give up electrons to become hydrogen ions. The hydrogen ions migrate through the electrolyte to the vicinity of the oxygen electrode, or cathode, while the loose electrons travel in an external circuit to the same electrode. At the cathode, oxygen, hydrogen ions, and electrons join to form water. The energy of this reaction is responsible for driving an electrical current through the external circuit. Each fuel cell produces about one volt, and many cells are joined in stacks to generate useful power. Modern cells usually operate at 250 °C.

The electrodes of a fuel cell consist of a porous, electrically conducting support, often made of carbon, filled with a well-dispersed catalytic component, usually fine grains of platinum powder. Without a catalyst, the chemistry at the site of oxygen gas consumption would take place too slowly for the cell to generate a significant current in the external circuit. Although increasing the amount of platinum increases the reaction efficiency, fuel-cell manufacturers find it uneconomical to use large quantities of platinum because it is so expensive. What they would like is an alternative electrode catalyst that is more efficient and less expensive than platinum.

The consequence of catalyst limitations is that the hydrogen fuel cell's realized efficiency is much lower that its theoretical efficiency; this consequence limits its practical utility. The key to achieving maximum efficiency is to find a material that effectively coordinates the interactions of four electrons, four hydrogen ions, and two oxygen atoms to produce two water molecules. The trickiest part of the task is to persuade an oxygen molecule to break up into two atoms and pick up four electrons. Until recently, no synthetic catalyst for this four-electron process existed. In 1983, researchers at Caltech and Stanford University (*2*) showed that it is possible to make a molecular catalyst for the four-electron reduction of oxygen and to confine it to an electrode. Their discovery has spurred a renewed effort to develop practical catalysts for the oxygen chemistry in a fuel cell.

Catalytic chemists are also challenged by the desire to use a liquid fuel, such as methanol, in a fuel cell. A liquid fuel would provide a system with more energy per unit volume, and it might be more useful in automobiles than the hydrogen-consuming devices developed so far. The practical rewards from a liquid fuel system are apparent, but the methanol chemistry in such a fuel cell is problematic. The desired process is one that involves six electrons per molecule of methanol. So far, no good catalyst exists for initiating such a process. However, scientists are studying a class of metals that may yet make such a catalyst possible.

Light Energy

After decades of trial and error, many researchers believe that *photovoltaic cells,* which are made of thin layers of silicon or other semiconducting materials and convert light directly into electric current, are rapidly approaching the time when they will be ready for widespread use. So far, the application of photovoltaic cells has been confined to power sources for spacecraft and remote communications devices. They also show up in consumer products, including calculators, portable radios, and watches. Cell efficiencies have improved to the point where solar cell modules are now a competitive energy source for users not connected to a utility grid. Some optimistic forecasters predict that by the end of the century, photovoltaic cells will be a cheaper source of electricity than oil.

A solid-state solar cell, such as the silicon-based photocells used today, is not the only device that converts sunlight directly into electricity. One alternative is the photoelectrochemical cell described earlier, which also allows the synthesis of various chemical products. Photoelectrochemical cells are potentially more versatile than solid-state photovoltaic devices because they can store solar energy by generating chemical fuels, in addition to producing electricity. For example, such cells, by light-driven chemical reactions, could convert inexpensive materials such as water and carbon dioxide into products such as hydrogen and methane, which can be burned, or used later to generate electricity in fuel cells. Photoelectrochemical cells also tend to be cheaper to make than solid-state photovoltaic cells.

A photoelectrochemical cell works because shining light on a semiconductor causes it either to contribute electrons to or to extract them from molecules in the liquid surrounding it. Thus the semiconductor develops an excess of electrons or a deficit of electrons (equivalent to positively charged "holes"), depending on whether it gains or loses electrons. This charge imbalance generates a current in an external circuit.

The major problem associated with such cells has been that of photocorrosion, a kind of light-induced rusting, that leads to decomposition of the semiconductor material. The very chemical reactions that are promoted by the light-driven semiconductor can destroy it. Another concern has been the low conversion efficiencies so far reported for these devices. The junction between the semiconductor and the electrolyte is much less efficient at separating electrons and holes than a photovoltaic p–n junction. Because the separation is not complete, the electrons and holes generated in a typical photoelectrochemical cell recombine, and the overall performance of the cell is poor. For liquid-junction cells to be technologically feasible, their stability and efficiency must be improved. If scientists and engineers succeed in overcoming these problems, photoelectrochemical cells may become a viable alternative to traditional energy sources.

Although the technology of semiconductor-liquid junction cells still lags substantially behind that of solid-state systems, progress in the technology is now rapid. Advances in suppressing photocorrosion of the light-sensitive electrode and in improving the cell's efficiency have come from understanding and controlling the interface between the electrodes and the electrolyte solution. In fuel-forming photoelectrochemical devices, the production of chemicals at the interfaces requires catalysis. Since the first demonstration of a durable, visible light-sensitive device in 1976 (3), dramatic progress has been made in raising the efficiency for solar conversion with such devices. Most recently, chemist Nathan Lewis of Stanford and his collaborators have reported (4) a photoelectrochemical device with a solar efficiency of 15 percent for the generation of electricity.

Lewis's new cell uses the semiconductor gallium arsenide. To minimize the problem of recombination, Lewis modified the surface of the semiconductor by exposing it to a solution of osmium trichloride for one minute. The osmium ions are absorbed into the surface of the gallium arsenide, where they hasten the transfer of "holes" into the electrolyte solution (a caustic solution of selenides). The osmium ions also help the semiconductor to catalyze the removal of electrons (oxidation) from the electrolyte. The combination of electrolyte oxidation and modification of the surface of the gallium arsenide is the key to the cell's improved performance.

Another significant advance in the technology of photoelectrochemical cells, reported in 1987 by Stuart Licht and his collaborators at the Weizmann Institute of Science in Israel (5), was the creation of a single device that combines photoelectrochemical conversion with electrochemical storage. In Licht's cell, the light-absorbing electrode is a single crystal of the semiconductor cadmium selenide telluride. The electrolyte is an aqueous polysulfide solution; a cobalt sulfide counterelectrode completes the circuit. In parallel with this electrode arrangement and connected by a permeable membrane is another cell containing an electrode composed of a combination of tin and tin sulfide. This cell is the storage part of the device. When light shines on the cadmium selenide telluride semiconductor, the photoelectrochemical half of the device produces more than a volt of electrical potential with a respectable 11.8 percent solar conversion efficiency. As this is happening, the storage half is also being charged through the tin sulfide electrode. In darkness, or below a certain level of light, the storage half of the cell begins delivering power from the tin sulfide electrode. With its own built-in storage, the cell could conceivably function at night, drawing on the energy stored during the day. The net result is that the cell continues to work, with an overall efficiency of 11.3 percent, regardless of the level of light.

Imitating Nature

Sunlight has more than enough energy to break the hydrogen–oxygen bonds in molecules of water. The problem is getting enough photons of light energy into the water molecules quickly enough to produce oxygen. Nature's answer is photosynthesis. Not only does photosynthesis supply oxygen by splitting water, it also provides energy for the plant. It does this by trapping the energy contained in sunlight and using it to convert carbon dioxide and water into oxygen gas and energy-rich substances called carbohydrates.

One promising approach for achieving highly efficient solar conversion at a low cost is to model the natural photosynthetic apparatus with molecular materials that absorb sunlight, separate charge, and catalyze fuel-forming reactions. These three functions of the natural photosynthetic apparatus occur with molecular materials, properly structured, in interfacial systems. Enough is now known about charge transport, catalysis, and light absorption to tackle the synthesis of rugged, efficient molecular assembles for solar conversion to electricity or to fuels.

To break down water molecules, the plant, or an artificial cell imitating the process, must capture the sun's light long enough to put it to work. Water itself is largely transparent to sunlight: the energy passes straight through it and very little is absorbed. Plants use green chlorophyll as their energy-trapping pigment. Researchers working on artificial photosynthesis are studying pigments that are called porphyrins, synthetic cousins of chlorophyll and hemoglobin (the pigment that gives red blood cells their color). When sunlight strikes porphyrins, the pigments release the electrons needed by the catalyst that kicks off the water-splitting process. The catalyst harnesses the released electrons to do the actual splitting of water into hydrogen and oxygen gas. Much of the research into artificial photosynthesis is now focused on devising better catalysts. The most commonly used catalysts to date, based on platinum, are both expensive and difficult to prepare.

Scientists have come up with experimental catalysts for photosynthesis-like systems that dispense with platinum altogether. One group of researchers uses particles of cadmium sulfide (doped with tiny amounts of ruthenium dioxide) immersed in hydrogen sulfide saturated water. In visible light, the system produces hydrogen and sulfur. Another possibility is to use ruthenium oxide in a colloidal form—a collection of very fine particles—as a catalyst. Much remains to be done in the science of solar energy conversion, but the overall efficiencies are now sufficiently high that it is difficult to rule out a technology based on light-induced oxidation and reduction chemistry at interfaces.

Exploiting Superconductors

In the past several months, scientists have made astonishing advances in the development of superconducting materials. The temperature at which superconductivity can be achieved has now been raised to more than 90 K, after years of stagnation at about 23 K. The advances in superconductor materials are truly revolutionary and could have important technological consequences for energy production and conversion. Long-distance electrical power transmission, levitated trains that ride on a magnetic field, powerful magnets for nuclear-fusion reactors or for medical imaging machines, and energy storage may all eventually benefit from the recent discoveries. In the near term, superconductor technology will facilitate the development of magnetic shielding, detectors of magnetic fields and infrared radiation, high-speed electronic signal processing, and voltage standards.

Large-scale exploitation of new superconductor materials will depend on a number of factors, many of which raise issues that will have to be addressed by chemists. Chemists will need to explore new methods of synthesizing materials. They will need to understand the chemistry that occurs during processing and be able to characterize superconducting materials with respect to structure and composition. They will need to be able to correlate structure and composition with functional properties, such as critical temperature, magnetic field, and current. One major challenge already evident is the need to understand the very nature of the high-temperature superconductor materials that have already been discovered. Advanced structural methods are likely to play a crucial role in the development of the necessary understanding.

One major challenge already evident is the need to understand the very nature of the high-temperature superconductor materials...

Putting the Squeeze on Oil

Interface science also plays a role in squeezing hydrocarbons out of rock, either by helping to increase the amount of crude oil that can be recovered from producing oil wells or by making it easier to extract oil from tar sands and shale. When an oil well is first drilled, the oil is pushed to the surface by the pressure of the gas–oil mixtures within the reservoir. As the gas and oil run out, the pressure decreases, and less oil flows to the surface. At that stage, the well can be flooded with water, sometimes containing additives such as surfactants similar to those found in detergents, to displace more oil. But even then, as much as two-thirds of the original oil is still underground, often trapped in minute pores in the rock.

For the last few decades, scientists and engineers have been developing "enhanced oil recovery" techniques for recovering hitherto unavailable reserves of oil. One approach is to inject a heat source such as steam into the reservoir to lower the viscosity of thick oils so that they can be pumped more readily to the surface. Another is to inject solvents such as carbon dioxide or an alcohol that dissolves hydrocarbons. The oil-carrying solvents can then be forced to the surface by displacement. Some researchers are also studying the use of water thickened with special polymers. The water–polymer mixture can efficiently sweep through a reservoir and push oil ahead of it. Researchers have also looked at ways of increasing the flow of natural gas through shales and other rocks.

Success in processing oil shale and tar sands also depends on good chemical engineering. Typically, the oil is obtained by simply heating crushed shale, either underground or at the surface, to convert the organic material it contains into oil. Much research has concentrated on how to make heating, or retorting, technology more efficient with fewer environmentally undesirable by-products. Understanding the kinetics, mechanisms, and reaction chemistry associated with various retorting technologies could lead to designs that can be applied to large shale deposits in environmentally acceptable ways.

Chemistry for Energy

The importance of energy to the nation's well-being is a great incentive and justification for a continuous, significant effort in energy research. Exciting progress in several areas of materials chemistry related to existing and emerging energy technologies has been made in very recent times. The fundamental advances now possible in the molecular-level understanding of interfaces promise to yield a rich knowledge base for future energy technologies.

References

1. *Opportunities in Chemistry*, National Academy of Sciences: Washington, DC, 1985.
2. Durand, R. R., Jr.; Bencosme, C. S.; Collman, J. P.; Anson, F. C. *J. Am. Chem. Soc.,* **1983,** *105,* 2710.
3. Ellis, A. B.; Kaiser, S. W.; Wrighton, M. S. *J. Am. Chem. Soc.* **1976,** *98,* 1635.
4. Tufts, B. J.; Abrahams, I. L.; Santangelo, P. G.; Ryba, G. N.; Casagrande, L. G.; Lewis, N. S. *Nature,* **1987,** *326,* 861.
5. Licht, S.; Hodes, G.; Tenne, R.; Manassen, J. *Nature,* **1987,** *326,* 863.

Chapter 11

High-Strength Composites

James Economy

In late 1986, an unusual, spindly aircraft named *Voyager* completed the first nonstop flight around the world without refueling. Although the flight represented no single, major technological breakthrough, it was the culmination of recent advances in engine and airfoil design and in the use of composite materials. The core of the featherweight plane's self-supporting skin, sandwiched between layers of carbon-fiber tape impregnated with an epoxy resin, consisted of a honeycombed, paperlike material threaded with tough fibers. Carbon-epoxy tubes supported its ungainly wings, giving the plane a wingspan longer than that of a Boeing 727. The only metal parts were its two engines and a few nuts and bolts. When empty, the no-frills aircraft weighed less than 2000 pounds.

Composite Materials

The *Voyager* aircraft is but one example of the progress that has been made over the last two decades in developing high-strength, lightweight composite materials to replace metals in structural applications. Composites now show up in sporting equipment, such as tennis rackets, golf clubs, baseball bats, skis, and bicycles, and also as automobile bodies and in military and civilian aircraft—wherever stiffness and strength must be combined with a low density. The U.S. Air Force's F-18 fighter jets are 10% composite, and the new *Stealth* bomber is said to be predominantly composite. Special composites, particularly those that can withstand high temperatures, have played important roles in the development of rockets and spacecraft.

In general, a *composite material* consists of particles or fibers of a reinforcing material embedded in some kind of matrix. Perhaps one of the earliest examples of a composite is the straw-strengthened mud used for construction in many cultures. Epoxy-reinforced glass fiber and Bakelite (a trademarked name for various plastics), which are composites, have been used for decades in a large number of applications, from insulation to cookware. Today, most widely used composites are resins with reinforcing fiber materials.

Although the field of high-performance or advanced composites is only 25 years old, tremendous progress has been made since the original work on new fibers began. Chemistry has played a pivotal role in the development of composites and advancements in the technology. We have arrived at the present level of sophistication primarily through the ingenuity of the chemists who designed processes to prepare both the stiff, high-strength fibers and the resins such as epoxies.

One of the first high-modulus, fibrous materials to be used in structural applications was the element boron. Boron fibers, however, were difficult to manufacture, tough to work with, and costly. The accelerated growth of advanced composite use required the development of carbon and aramid fibers on a commercial scale in the 1960s.

Aramid (or Kevlar) fibers, which are made up of long chains of linked hexagons of carbon atoms, tend to be more tough and somewhat less stiff than graphite fibers. Because they are relatively easy to manufacture, carbon and aramid fibers, used separately or together, are the work horses of advanced composite structural design and manufacturing.

Fabrication

A typical carbon-based composite begins as filaments of polyacrylonitrile fiber, which is a cousin to rayon and similar to the acrylic fiber that goes into sweaters and other knit clothing. Each fiber is considerably thinner than a human hair. Stretching the fibers, then roasting them at a very high temperature in the absence of oxygen leaves behind fiber with very high strength. Sometimes the carbon fibers are later heated to even higher temperatures to increase their strength further. The fibers are then processed into tapes, yarns or woven fabric, or they are chopped up as filler. A one-inch-wide strip of carbon-fiber tape may contain 40,000 or more individual carbon fibers.

If left unsupported, the carbon fibers prove to be stiff and somewhat brittle. Embedding them in a polymer endows the resulting material with the immense strength of the carbon fiber. The composite's mechanical properties are governed not by the fibers alone, but by a synergy between the fibers and the matrix. The matrix itself acts as an adhesive, binding the fibers and lending a solidity to the material. It protects the fibers from environmental stress and physical damage that could initiate cracks. Although polymer matrices are most common, carbon fibers can also be embedded in glass, ceramics, and even metals, depending on the requirements of any particular application. Fiberglas is a composite composed of short glass fibers embedded in polyester.

In the fabrication of a typical polymer–matrix composite, the fibers usually begin in the form of yarns or bundles. They are impregnated with the matrix resin and can then be formed into tapes and sheets or woven into fabrics, which are often assembled by hand into a laminated structure. Filaments can also be wound into the shapes of large structures, including the fuselages of light airplanes. The filament shapes are thoroughly soaked with epoxy resin and then cured. Such a process, though labor-intensive, eliminates the need for massive cutting, bending, stamping, forming, and milling equipment, which is necessary for metal work.

Properties

A composite material's most attractive property is its unusually high strength-to-weight ratio. Structural components made of carbon fiber are every bit as strong as their steel counterparts but weigh 50 to 70% less. Similarly, composites have a superior stiffness-to-weight ratio. In engineering parlance, such stiff materials have a high modulus. They also have excellent resistance to fatigue and corrosion, and expand very little as temperature increases. In fact, some composites have a negative coefficient of thermal expansion, which means that unlike most materials, the composite actually shrinks a little when it warms up. Furthermore, composites are relatively easy to mold into parts with complex shapes. The fabrication of similar parts from metal would require the machining of many separate components that would have to be assembled. All of these advantages are now well understood, and structural engineers routinely take such properties into account when designing structural parts. Composite materials make it feasible to design and build structures that otherwise would be impossible to realize.

Nevertheless, composites are not ideal materials. They have their own particular weaknesses, faults, and disadvantages. Although they are strong, composites are easy to damage. It is practically impossible to drill a hole into a composite material without causing splintering, and laminated structures may, under extreme conditions, find their layers peeling apart. Many composites cannot survive high temperatures. Epoxy, for example, begins to decompose or soften below 180 °C. Composites are also expensive to manufacture. A carbon–epoxy composite may cost $20 per pound compared with a cost of $2.50 per pound for aluminum.

Engineers and designers would like to have cheaper, even stronger, more heat-resistant composites that are also easy to work with. Today, scientists are on the verge of a number of major advances in producing composite materials that could greatly increase the performance characteristics of composites and permit a much broader use of this unique class of materials. The advances are the result of a number of critical breakthroughs in the synthesis of new reinforcing agents and matrices.

Composite materials make it feasible to design and build structures that otherwise would be impossible to realize.

Much of the latest progress can be traced to an increasingly sophisticated understanding of the molecular and morphological features of the matrix and the reinforcing agent. This understanding is likely to result in new fiber–matrix combinations that can be widely used in aircraft and automobile bodies. It should also make possible composite ceramics suitable for use in engines, and lower cost molecular composites that can have properties tailored to specific applications.

New Directions for Higher Strength Carbon Fibers

For some time, scientists have known that the strength and other properties of acrylic-based carbon fibers vary across the diameter of a fiber. The fiber's strength is much higher at its surface than at its core. The cause of this gradation is generally attributed to the shear forces that had been exerted on the fibers when they were drawn into fine filaments. In a sense, the fiber surfaces are better aligned than they were before the drawing process. With this knowledge, fiber scientists have been gradually reducing the fiber diameter from 9 to 7 micrometers, and now to 5 micrometers, to achieve a much higher surface-to-core ratio and improved mechanical properties.

Today, carbon and graphite fibers with incredibly high strength and stiffness have been developed, along with extremely tough polymeric matrices that greatly enhance damage tolerance.

Improved epoxy resins and polyimide resins that can survive continuous exposure to temperatures of more than 250 °C have been developed to take better advantage of the greatly improved reinforcing potential afforded by the new fibers. The new, tough, polymeric matrices greatly extend the potential uses of graphite-fiber composites in load-bearing applications in commercial aircraft.

Planar Reinforcement

One of the most fascinating materials with potential as a reinforcing agent is aluminum diboride. It can be produced as single-crystal flakes that measure almost half a centimeter wide and only several micrometers thick (*see* Figure 1). Such a structure could allow scientists to create new materials that display great strength in two dimensions, similar to the strength that graphite fibers show in one dimension.

In the early 1970s, aluminum diboride flakes were

Figure 1. AlB_2 single crystal flakes.

incorporated into composites that had planar stiffness values of 40 million pounds per square inch. That value compares favorably with the 10 million pounds per square inch that typically can be attained with graphite fibers that are cross-plied for planar reinforcing. The flakes were also easy to process and had excellent damage tolerance. One can even drill holes into the material for attachments. The key drawback was the composite material's limited ability to withstand stretching and the application of pressure.

Now scientists have developed new tougher polymeric matrices in which aluminum boride flakes can be embedded. They will be able to begin taking advantage of the added planar strength afforded by the flakes. With tougher matrices, it would be possible to obtain a much higher planar strength in the composites, while permitting a significant increase in design strain. Flake-reinforced composites would probably cost about as much as typical graphite fibers. Furthermore, because aluminum diboride flakes are made from an aluminum melt, they open the opportunity for fabricating aluminum-based composites directly.

Ceramic Composites

Fiber-reinforced composites are not always made with polymeric resins. Some extremely high-temperature applications require ceramic materials. Significant progress has been made in developing inorganic, or ceramic, composites for use at temperatures of more than 1000 °C, a goal critical to the development of advanced ceramic engines and vehicles designed to travel at speeds greater than three times the speed of sound.

The problem with almost all ceramics is their intrinsic brittleness and resultant low reliability. Dropping a metal cup merely dents it, whereas a typical ceramic cup is shattered. Presumably, a fiber-reinforced ceramic composite would greatly reduce the brittleness and provide much greater fatigue resistance. The fibers in a ceramic-matrix composite fill that need by blocking the growth of cracks. A growing crack that encounters bits of fiber may either be deflected, or be made to pull the fiber from the matrix. In either case, the process would absorb energy and slow the growth of the crack. Even if the matrix were to become riddled with cracks, it would fracture less readily than an unreinforced ceramic because of the many fibers that bridge the cracks.

In fact, the possibility of manufacturing tough ceramic composites able to withstand high temperatures has already been demonstrated with carbon-fiber-reinforced carbon-matrix composites, which have been available since the late 1960s. Such carbon–carbon composites (*see* Figure 2) have been used in aircraft brakes, rocket motors, missile components, and spacecraft. These composites are ductile and show none of the traditional brittle behavior associated with solid bodies of carbon and graphite. Unfortunately, they are stable at 2000 °C only under inert conditions and tend to oxidize rapidly in air at temperatures as low as 400 °C. A thin layer of ceramic is often applied to the surface of a carbon–carbon composite to protect it.

Figure 2. Carbon–carbon composites.

Stringent requirements for the fiber and matrix necessary to achieve the desired thermal and mechanical composite properties limit the number of candidates. In fact, the only candidate fibers with acceptable mechanical properties and thermal stabilities are boron carbide and silicon carbide, with densities of 2.50 and 3.25 grams per cubic centimeter, respectively. In the selection of a suitable matrix, one must pick not only a material that is stable in air at 1100 °C, but also one that lends itself to processing in the form of a laminate. Because a process for fabricating composites with a carbon matrix has been developed for carbon–carbon composites, it would seem reasonable first to prepare the ceramic composite with a carbon matrix, and then to convert the carbon matrix to silicon carbide by silicon infiltration, or to protect the carbon with a chemical-vapor-deposited coating of silicon carbide.

The feasibility of such an approach has already been demonstrated using boron carbide fibers in a carbon matrix coated with silicon carbide. Compared with a carbon–carbon composite, the boron carbide fiber composite shows a sevenfold improvement in stiffness, a twofold increase in strength, and a fivefold increase in interlaminar shear strength. In fact, this approach seems to satisfy almost all of the required properties, including stability in air at 1100 °C.

The boron carbide fibers (*see* Figure 3) are prepared by a novel process in which a continuous filament carbon yarn is chemically converted to boron carbide. The reaction, which occurs at 1800 °C, is very rapid and is completed within 15–30 seconds. But boron carbide yarn is likely to be relatively expensive, because it requires the use of carbon yarn, and that alone costs $20 per pound.

Figure 3. B$_4$ fibers.

The ability to convert a carbon fiber chemically to boron carbide is unusual because one would normally expect to coat the carbon rather than to convert it. The crystal structure of boron carbide consists of large, close-packed, quasi-spherical units of 12 boron atoms. The boron atoms, derived from boron chloride, are able to diffuse easily through the large interstitial openings made by carbon that has already been converted to boron carbide. The boron atoms react with the carbon core that still remains.

Researchers have experimented with a wide range of ceramic composites, with varying results. United Technologies scientists, for example, several years ago developed a lithium aluminosilicate glass-ceramic reinforced with silicon carbide fibers. The composite remained resistant to fracture at temperatures as high as 1000 °C. That makes it usable in gas turbine engines. Meanwhile, scientists at General Electric were working on a composite composed of elemental silicon reinforced with silicon-carbide fibers formed when carbon fibers are exposed to molten silicon. The silicon reacts with the carbon to form, in a matrix of silicon, lines of silicon carbide crystals in the shape of the original carbon fibers. This composite has been used as a liner for a combustion chamber.

Ordered Polymers

Organic Polymer Fibers

The notion that many materials perform best when made into fibers also holds for certain organic polymers. Polymer molecules consist of long chains of atoms, usually carbon, joined by covalent bonds. Under most conditions, the chains are either loosely tangled or crystallized in complex patterns. They can be pulled apart fairly easily, and as a result, the bulk material is flexible and weak. If the chains are all oriented in the direction of stress, however, the polymer can be very strong and stiff. Certain polymer molecules are rodlike, and they readily become aligned when the polymer is spun into a fiber. The production of aramid fibers, which are notable for their strength and stiffness, is based on this effect. Polymers with flexible chains, such as polyethylene, can also be made into extremely strong and stiff fibers through new techniques that extend and orient the polymer molecules along the fiber axis.

Rodlike polymers display a propensity to form molecular composites. The concept of molecular composites has been known for more than 15 years, but only in the past few years have scientists made meaningful progress in developing ones with desirable properties. Some of the best examples of molecular composites fall into the category of liquid crystal polymers.

Liquid Crystal Polymers

Liquid crystal materials, now familiar for their use in

displays for digital watches and calculators, consist of relatively small molecules that have the unusual property of being able to arrange themselves into highly ordered systems. The result is a state of matter that is more structured than a liquid but less so than a solid crystal. Such materials possess characteristics of both liquids and solids. (*See* Figure 4.)

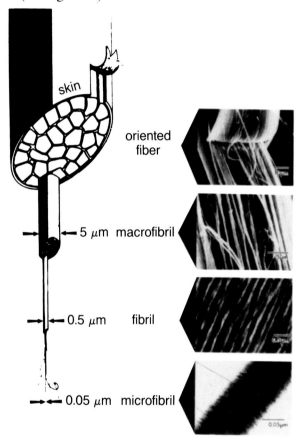

Figure 4. Structural model of a liquid crystal polymer.

Liquid crystal polymers consist of relatively large molecules, which organize themselves when melted, and retain that structure even when subsequently solidified. They are generally solids at room temperature and have some of the desirable properties of plastics, ceramics, and metals. Liquid crystal polymers are chemically stable, can withstand high temperatures, and expand very little when heated. They are lightweight but very strong. That increase in strength can be attributed to the high degree of orientation present when the polymers are processed in the liquid crystal state. Like most polymers, they are relatively easy and inexpensive to process. Fibers made from liquid crystal polymers are tough and chemically stable. In fact, the aramid fiber Kevlar is itself a liquid crystal polymer. The first applications of liquid crystal polymers with melting points above 400 °C was oven-proof plasticware—plastic containers in which food could be cooked in a conventional oven.

To date, most of the interest in liquid crystal polymers has been in making use of their high degree of organization in the melt or solution state for processing polymers into highly ordered materials with enhanced mechanical properties. One exciting endeavor is the recent development of self-reinforcing liquid crystalline polyesters, which can be injection molded into shapes with a strength and stiffness approaching that of aluminum. The potential low cost of these polyesters and the ease of processing them into shapes suggest that such materials could be used in structural parts for automobiles.

Examining liquid crystalline polyesters under a microscope reveals their highly ordered, fibrous nature. Moreover, each fiber shows a complex hierarchy of reinforcing fibrils, starting with fibrils that are barely larger than typical polymer molecules and ranging up to fibrils 5 micrometers in diameter. The presence of this type of structure has led to the use of the term "self-reinforcing" to describe these plastics.

Improving Mechanical Properties

The potential for improved mechanical properties through formation of molecular composites is demonstrated when minor orientation from injection molding yields specimens with greatly increased mechanical properties in the direction of flow. The molded objects have mechanical properties in the direction in which they were stretched that are more than tenfold greater than those in the cross direction. The process of injection molding itself may increase a sample's tensile strength from 900,000 to 2.7 million pounds per square inch. With higher orientation such as that observed in fiber forming, much higher mechanical properties can be achieved. Relying on achieving a high degree of orientation in its fibers, Sumitomo Chemical announced in 1985 the availability of an Ekonol fiber with reported properties significantly better than Kevlar. Of the possible molecular composites, the most exciting are the melt-processible aromatic polyesters. Today, three such copolyesters are commercially available.

To exploit the full potential of molecular composites, scientists must design low-cost injection molding techniques that produce shapes with the desired orientation. One approach described recently depends on rotational injection molding to produce shapes that display high radial and circumferential strength values.

Composites for the Future

Tremendous opportunities now exist for developing composites with greatly improved strength and damage tolerance, and also for beginning to develop high-temperature ceramic composites for prototype evaluation. Furthermore, the unique properties of liquid crystal polymers can now be exploited to create entirely new materials. Composite use has grown exponentially over the last decade. Whereas carbon fibers were available only in laboratory quantities in 1965, the Boeing 767 and 757 alone now account for deliveries of more than 10,000 pounds of carbon-fiber material per month.

The approximate market size for high-performance composites and its projected growth can be estimated from the current and projected market for graphite fibers based on polyacrylonitrile. The current market for graphite fibers is about $80 million per year, but their high cost of $20 per pound clearly limits their use. Aerospace applications clearly dominate the market, and the aerospace industries' need for lightweight composites will undoubtedly continue to grow.

During the last three decades, work on advanced composite materials has led to some of the strongest, lightest, stiffest, most corrosion-resistant materials now available. The technology is no longer in its infancy. New fibers, new types of matrices including metals, and new lamination methods will undoubtedly in the future allow many applications that now seem impractical.

Index

Index

A

Acrylic-based carbon fibers, new directions, 120

Acquired immune deficiency disease, *See* AIDS

Active sites of enzymes, mutation, 50

Affinity cleaving, description, 24

Agricultural biotechnology, definition, 8

Agrobacterium tumefaciens, natural genetic engineer, 17

Agrochemicals, designing, 14

AIDS
  and biotechnology, 8, 37
  and human immunodeficiency virus, 38

Aluminum diboride, reinforcing agent, 121

Amino acids
  in proteins, 54
  manufacturing, 57

Animal toxicity, herbicides, 15, 16

Antibodies
  converting into enzymes, 58
  definition, 28
  engineering, 58
  function, 58

Anticancer activities, cisplatin and *trans*-DDP, 37

Antigen, definition, 28

Antiretroviral activity, AZT, 40

Antitrypsin, in human lung, 56

Antitumor drugs, interrupting DNA synthesis, 33

Antiviral drugs, interrupting DNA synthesis, 33

Arginine transaminase, converted from aspartate transaminase, 58

Aspartame, composition, 57

Aspartate transaminase, converted to arginine transaminase, 57

Atomic views, 89

Atoms, behavior, 73

Auger photoelectron spectroscopy, to study surface materials, 104

3'-Azido-3'-deoxythymidine, *See* AZT

AZT
  antiretroviral activity, 40
  structure, 40

B

Base pairs in DNA molecules, 21

Biological systems, evolution, 9

Biology and chemistry, convergence, 11

Biotechnology
  agricultural, definition, 8
  and AIDS, 8, 37
  applications in agriculture, 14–17
  applications in health sciences, 18
  applications to pharmaceutical products, 7
  basis, 12
  four chemical technologies, 49
  improving our lives, 7
  origins, 3
  second chemical revolution, 54
  significance for industry, 13

Boron carbide fiber composites, properties, 124

Boron carbide fibers, preparation, 124

C

Cadmium sulfide catalyst for artificial photosynthesis, 113

Cancer chemotherapy, 59

Carbon-based composites, fabrication, 118

Carbon–carbon composites, uses, 123

Catalysis
  and interfaces, 103
  heterogeneous, definition, 102

Catalysts
  definition, 75
  for artificial photosynthesis, 113
  mechanism of action, 102
  platinum, 77
  surface, 103
  zeolites, 76

Catalytic activity of enzymes, 50

Catalytic antibodies, development, 28

Ceramic composites, 122

Ceramic materials, properties, 79

Chemical-vapor deposition, 95

Chemical electricity, 106

Chemical kinetics in materials science, 66

Chemical reactions, methods of studying, 103

Chemical revolution, second, 53

Chemical technologies that make up biotechnology, 49

Chemical theory in materials science, 67

Chemist's approach to biotechnology, 3

Chemistry
  and biology, convergence, 11
  contribution to wedding of biology and chemistry, 9
  driving force in materials design, 73
  in materials science, 64

Book and cover design: Carla L. Clemens
Copy editor: Janet S. Dodd

Typesetting: Hot Type Ltd., Washington, DC
Printing: The Sheridan Press, Hanover, PA
Clothbound binding: Maple Press, Manchester, PA